MEMOIRS
of the
American Mathematical Society

ISSN 0065-9266 (print)
ISSN 1947-6221 (online)

Number 1230

Algebras of Singular Integral Operators with Kernels Controlled by Multiple Norms

Alexander Nagel
Fulvio Ricci
Elias M. Stein
Stephen Wainger

November 2018 • Volume 256 • Number 1230 (sixth of 6 numbers)

AMERICAN MATHEMATICAL SOCIETY

Library of Congress Cataloging-in-Publication Data

Cataloging-in-Publication Data has been applied for by the AMS. See http://www.loc.gov/publish/cip/.

DOI: https://doi.org/10.1090/memo/1230

Memoirs of the American Mathematical Society

This journal is devoted entirely to research in pure and applied mathematics.

Subscription information. Beginning with the January 2010 issue, *Memoirs* is accessible from www.ams.org/journals. The 2018 subscription begins with volume 251 and consists of six mailings, each containing one or more numbers. Subscription prices for 2018 are as follows: for paper delivery, US\$999 list, US\$799.20 institutional member; for electronic delivery, US\$879 list, US\$703.20 institutional member. Upon request, subscribers to paper delivery of this journal are also entitled to receive electronic delivery. If ordering the paper version, add US\$20 for delivery within the United States; US\$80 for outside the United States. Subscription renewals are subject to late fees. See www.ams.org/help-faq for more journal subscription information. Each number may be ordered separately; *please specify number* when ordering an individual number.

Back number information. For back issues see www.ams.org/backvols.

Subscriptions and orders should be addressed to the American Mathematical Society, P. O. Box 845904, Boston, MA 02284-5904 USA. *All orders must be accompanied by payment.* Other correspondence should be addressed to 201 Charles Street, Providence, RI 02904-2213 USA.

Memoirs of the American Mathematical Society (ISSN 0065-9266 (print); 1947-6221 (online)) is published bimonthly (each volume consisting usually of more than one number) by the American Mathematical Society at 201 Charles Street, Providence, RI 02904-2213 USA. Periodicals postage paid at Providence, RI. Postmaster: Send address changes to Memoirs, American Mathematical Society, 201 Charles Street, Providence, RI 02904-2213 USA.

This publication is indexed in *Mathematical Reviews®*, *Zentralblatt MATH*, *Science Citation Index®*, *Science Citation Index™-Expanded*, *ISI Alerting Services℠*, *SciSearch®*, *Research Alert®*, *CompuMath Citation Index®*, *Current Contents®/Physical, Chemical & Earth Sciences.*
This publication is archived in *Portico* and *CLOCKSS*.
Printed in the United States of America.

♾ The paper used in this book is acid-free and falls within the guidelines established to ensure permanence and durability.
Visit the AMS home page at https://www.ams.org/

10 9 8 7 6 5 4 3 2 1 23 22 21 20 19 18

Contents

Abstract

We study algebras of singular integral operators on $\mathbb{R}^n$ and nilpotent Lie groups that arise when considering the composition of Calderón-Zygmund operators with different homogeneities, such as operators occuring in sub-elliptic problems and those arising in elliptic problems. These algebras are characterized in a number of different but equivalent ways: in terms of kernel estimates and cancellation conditions, in terms of estimates of the symbol, and in terms of decompositions into dyadic sums of dilates of bump functions. The resulting operators are pseudo-local and bounded on L^p for $1 < p < \infty$. While the usual class of Calderón-Zygmund operators is invariant under a one-parameter family of dilations, the operators we study fall outside this class, and reflect a multi-parameter structure.

Received by the editor November 17, 2015 and, in revised form, April 8, 2016.

Article electronically published on September 24, 2018.

DOI: https://doi.org/10.1090/memo/1230

2010 *Mathematics Subject Classification.* 42B20, 35S05.

Key words and phrases. Singular integral operators, Multiplier operators, Multi-parameter theory, Pseudo-differential operators, Homogeneous nilpotent Lie groups, Flag kernels.

The first author is affiliated with the University of Wisconsin-Madison, Madison, Wisconsin 53706. Email: nagel@math.wisc.edu.

The second author is affiliated with the Scuola Normale Superiore, Piazza dei Cavalieri 7, 56126 Pisa, Italy. Email: fricci@sns.it.

The third author is affiliated with Princeton University, Princeton, New Jersey 08544. Email: stein@math.princeton.edu.

The fourth author is affiliated with the University of Wisconsin-Madison, Madison, Wisconsin 53706. Email: wainger@math.wisc.edu.

CHAPTER 1

Introduction

In this paper we study algebras of singular integral operators on $\mathbb{R}^n$ and nilpotent Lie groups that arise when considering the composition of Calderón-Zygmund operators with different homogeneities, such as operators occuring in sub-elliptic problems and those arising in elliptic problems. For example, one would like to describe the algebras containing the operators related to the Kohn-Laplacian for appropriate domains or those related to inverses of Hörmander sub-Laplacians, when these are composed with the more standard class of pseudo-differential operators. The algebras we study can be characterized in a number of different but equivalent ways, and consist of operators that are pseudo-local and bounded on L^p for $1 < p < \infty$. While the usual class of Calderón-Zygmund operators is invariant under a one-parameter family of dilations, the operators we study fall outside this class, and reflect a multi-parameter structure.

This paper is the second in the series begun with [**NRSW12**]. We thank Detlef Müller for his exceptionally careful reading of this manuscript and for his very useful comments and suggestions.

1.1. Background

An initial impetus for the study of composition of singular integral operators of different types came from the study of the $\overline{\partial}$-Neumann problem in complex analysis. In the case of domains in $\mathbb{C}^n$ where matters are sufficiently well understood, the relevant "Calderón operator" for this boundary-value problem can be viewed as a composition of a sub-elliptic type operator with a standard pseudo-differential operator of order 0. Early studies of such compositions in the context of the Heisenberg group can be found [**GS77**] and in [**PS82**]. In [**MRS95**] more general such operators appear as singular integrals with "flag kernels". The corresponding algebra was broad enough to contain, for example, all operators of the form $m(\mathcal{L}, iT)$, where $\mathcal{L}$ is the sub-Laplacian and T is the central invariant vector field, with m a multiplier of Marcinkiewicz-type. The notion of flag kernels (having singularities on appropriate flag varieties) and the properties of the corresponding singular integrals were then extended to the higher step case in [**NRS01**], largely in the Euclidean setting, and then in [**NRSW12**] in the context of automorphic flags on a general homogeneous group. In this connection we should mention the general theory of Brian Street [**Str14**] which treats aspects of both singular integrals and singular Radon transforms in the context of multi-parameter analysis.

However, the general nature of these results did not provide an entirely satisfactory answer to a main question that arises when composing singular integrals of different homogeneities: that of characterizing the resulting class of kernels. It turns out that the class in question is indeed narrower than the class of flag kernels: the kernels satisfy stricter differential inequalities, and in particular they are all smooth away from the origin, and the corresponding operators are thus pseudo-local.

A different approach to the study of flag kernels can be found in the work of Głowacki; see [**Gło10a**], [**Gło10b**], [**Gło13**]. There has been recent interest in the study of Hardy spaces associated to flags and flag kernels (see for example [**HL10**], [**HLL10**], [**Rua11**], [**HLL13**], [**HLS14**], [**Wu14a**]) and in the study of weighted norm inequalities (see for example [**Wu14b**]). Further recent references that deal with flag kernels include [**Yan09**], [**DLM10**] [**Rua10**], [**WL12**], [**LZ13**], [**SY13**], [**Wu14b**], and [**MPR15**].

1.2. Some motivating examples

We begin by describing a particularly simple situation occurring in studying the heat equation on $\mathbb{R}^n \times \mathbb{R}$, or in studying convolution of operators arising on the non-Abelian Heisenberg group $\mathbb{H}^n = \{(\mathbf{z}, t) : \mathbf{z} = (z_1, \ldots, z_n) \in \mathbb{C}^n,\ t \in \mathbb{R}\}$. In the latter case, Calderón-Zygmund kernels adapted to the automorphic dilations $\delta \cdot (\mathbf{z}, t) = (\delta \mathbf{z}, \delta^2 t)$ are distributions $\mathcal{K}$ on $\mathbb{H}^n$ which, away from the origin, are given by integration against a smooth function K satisfying

$$\left|\partial_t^\beta \partial_{z,\bar{z}}^\alpha K(\mathbf{z}, t)\right| \lesssim \left(|\mathbf{z}| + |t|^{\frac{1}{2}}\right)^{-2n-2-|\alpha|-2\beta} \tag{1.1}$$

and which satisfy appropriate cancellation conditions. Similarly the isotropic variants of these kernels (the building blocks of the standard pseudodifferential operators) satisfy instead the inequalities

$$\left|\partial_t^\beta \partial_{z,\bar{z}}^\alpha K(\mathbf{z}, t)\right| \lesssim (|\mathbf{z}| + |t|)^{-2n-1-|\alpha|-\beta}. \tag{1.2}$$

If we consider kernels with compact support, it turns out that the relevant algebra of kernels containing both kinds of distributions are those given by the mixed differential inequalities

$$\left|\partial_t^\beta \partial_{z,\bar{z}}^\alpha K(\mathbf{z}, t)\right| \lesssim (|\mathbf{z}| + |t|)^{-2n-|\alpha|} \left(|\mathbf{z}|^2 + |t|\right)^{-1-\beta} \tag{1.3}$$

for $(\mathbf{z}, t)$ in the unit ball, in addition to cancellation conditions.

There are several ways of thinking about the estimates in equation (1.3). One notices that the inequalities in (1.3) are the best that can be satisfied on the unit ball for kernels K that are assumed to be either of type (1.1) or of type (1.2), but this rather simple situation is not repeated in the higher step case[1]. However, one can make two more productive observations.

a) Operators satisfying (1.3) can be understood in terms of the theory of *flag kernels* studied in [**NRS01**] and [**NRSW12**]. Consider the following two sets of differential inequalities:

$$\left|\partial_t^\beta \partial_{z,\bar{z}}^\alpha K(\mathbf{z}, t)\right| \lesssim |\mathbf{z}|^{-2n-|\alpha|} \left(|\mathbf{z}|^2 + |t|\right)^{-1-\beta}, \tag{1.4}$$

$$\left|\partial_t^\beta \partial_{z,\bar{z}}^\alpha K(\mathbf{z}, t)\right| \lesssim \left(|\mathbf{z}| + |t|\right)^{-2n-|\alpha|} |t|^{-1-\beta}. \tag{1.5}$$

Equation (1.4) gives the differential inequalities satisfied by flag kernels for the flag $(0) \subset \mathbb{C}^n \subset \mathbb{C}^n \oplus \mathbb{R}$, while equation (1.5) give the differential

[1]See section 1.3.8 below for an interesting example, and Chapter 7 for a more complete discussion

inequalities for the opposite flag $(0) \subset \mathbb{R} \subset \mathbb{C}^n \oplus \mathbb{R}$. Operators of type (1.1) satisfy (1.4), operators of type (1.2) satisfy (1.5), and locally (for $(\mathbf{z}, t)$ in compact sets), operators of type (1.1) or (1.2) satisfy *both* (1.4) and (1.5). Moreover, we will see that a kernel satisfying both (1.4) and (1.5) automatically satisfies (1.3). Thus one is led to the study of *two-flag kernels* which locally are simultaneously flag kernels for two opposite flags.

b) A second related perspective is to view kernels K satisfying the conditions of (1.3) as satisfying differential inequalities appropriate to two different families of dilations on $\mathbb{R}^N$. Derivatives in z or $\bar{z}$ are controlled by the dilations $(\mathbf{z}, t) \to (\delta\mathbf{z}, \delta t)$, while derivatives with respect to t are controlled by the dilations $(\mathbf{z}, t) \to (\delta\mathbf{z}, \delta^2 t)$. This suggests that, more generally, we should study operators whose kernels satisfy differential inequalities in which different derivatives are controlled by different families of dilations.

In this paper we begin by adopting the second point of view. It will then turn out that the resulting class of distributions includes the class which belong to two flags. One of the main goals of this paper is the study the composition of convolution operators with kernels of this form. In the Euclidean context this reduces to the study of the product of the Fourier multipliers, but in the non-Abelian situation the composition becomes a more serious issue.

A second goal of this paper is to establish regularity of two-flag kernels. The inequalities in equations (1.4) and (1.5) are the differential inequalities for a kernel belonging simultaneously to two opposite 2-step flags, and it is easy to see that these imply the inequalities in equation (1.3). However in the case of higher step, the results are not so simple, and in fact are rather surprising. Consider, for example, the case of two opposite 3-step flags in $\mathbb{R}^3$:

$$\begin{aligned} \mathcal{F}: &\quad (0) \subset \{(x,y,z) : y = z = 0\} \subset \{(x,y,z) : z = 0\} \subset \mathbb{R}^3, \\ \mathcal{F}^\perp: &\quad (0) \subset \{(x,y,z) : x = y = 0\} \subset \{(x,y,z) : x = 0\} \subset \mathbb{R}^3. \end{aligned}$$

A flag kernel $\mathcal{K}$ for the flag $\mathcal{F}$ with homogeneity $\lambda \cdot (x,y,z) = (\lambda x, \lambda y, \lambda z)$ satisfies the differential inequalities

$$\big|\partial_x^a \partial_y^b \partial_z^c K(x,y,z)\big| \le C_{a,b,c}\, |x|^{-1-a}(|x| + |y|)^{-1-b}(|x| + |y| + |z|)^{-1-c}. \tag{1.6}$$

A flag kernel $\mathcal{K}$ for the flag $\mathcal{F}^\perp$ with homogeneity $\lambda \cdot (x,y,z) = (\lambda x, \lambda^{\frac{1}{2}} y, \lambda^{\frac{1}{3}} z)$ satisfies the differential inequalities

$$\big|\partial_x^a \partial_y^b \partial_z^c K(x,y,z)\big| \le C_{a,b,c}\, (|x| + |y|^2 + |z|^3)^{-1-a}(|y| + |z|^{\frac{3}{2}})^{-1-b}|z|^{-1-c}. \tag{1.7}$$

Suppose that $\mathcal{K}$ is a distribution with compact support which is a flag kernel for both flags. Looking only at the inequalities in equations (1.6) and (1.7), these estimates provide no information about $K(x,y,z)$ if $x = z = 0$ and $y \neq 0$; in terms of the assumed size estimates alone, the distribution $\mathcal{K}$ could be singular away from the origin. Unexpectedly, it turns out that

$$\begin{aligned} &\big|\partial_x^a \partial_y^b \partial_z^c K(x,y,z)\big| \\ &\quad \le C_{a,b,c}\, (|x| + |y|^2 + |z|^3)^{-1-a}(|x| + |y| + |z|^{\frac{3}{2}})^{-1-b}(|x| + |y| + |z|)^{-1-c}, \end{aligned}$$

and the distribution $\mathcal{K}$ is indeed pseudo-local. It is important to note that this is a consequence of both the differential inequalities and the cancellation conditions.

1.3. Outline of results

Our results can be divided roughly into two kinds: those concerning properties of the kernels and those concerning the resulting convolution operators.

1. We begin by studying a class $\mathcal{P}(\mathbf{E})$ of distributions on $\mathbb{R}^N$ given away from the origin by integration against a smooth function. These functions are required to satisfy differential inequalities adapted to a set of $n \leq N$ families of dilations on $\mathbb{R}^N$ prescribed by an $n \times n$ matrix $\mathbf{E}$. Distributions $\mathcal{K} \in \mathcal{P}(\mathbf{E})$ must also satisfy appropriate cancellation conditions.
2. We are primarily interested in the local behavior of these distributions[2], and it will be convenient to modify the class of kernels outside the unit ball so that they and all their derivatives are rapidly decreasing at infinity. Under these hypotheses, we shall denote the class of modified kernels by $\mathcal{P}_0(\mathbf{E})$.
3. We then characterize these distributions in terms of their Fourier transform, and also in terms of decompositions as infinite dyadic sums of dilates of normalized bump functions. This material is discussed in Chapters 2 - 6. In Chapter 7 we show that these distributions fall outside the class of standard Calderón-Zygmund kernels. In Chapter 10 we find the smallest class $\mathcal{P}_0(\mathbf{E})$ containing the convolution of Calderón-Zygmund kernels with different homogeneities. In Chapter 11 we show that two-flag kernels provide examples of the classes $\mathcal{P}(\mathbf{E})$.

   After establishing the basic properties of the distributions $\mathcal{K} \in \mathcal{P}_0(\mathbf{E})$, we fix a homogeneous nilpotent Lie group G whose underlying space is $\mathbb{R}^N$ and thus has a given automorphic family of dilations. (Of course this includes the special case in which the group G is the abelian $\mathbb{R}^n$.)
4. If the n families of dilations on $\mathbb{R}^N$ given by the matrix $\mathbf{E}$ are compatible[3] with the automorphic dilations, we show that convolution on G with a distribution $\mathcal{K} \in \mathcal{P}(\mathbf{E})$ defines a bounded operator on $L^p(G)$ for $1 < p < \infty$, and that the collection of these convolution operators form an algebra under composition. This material is discussed in Chapters 8 - 9.
5. We also characterize the smallest algebra of convolution operators which arise when one composes Calderón-Zygmund operators of different homogeneities. This material appears in Chapter 10.
6. Finally we describe a generalization of the class $\mathcal{P}(\mathbf{E})$ to allow variable coefficients, and we investigate the commutation properties of these operators. In particular, we study the role of the classical pseudo-differential operators. This material is considered in Chapters 12 and 13.

In the following sections we provide a more detailed summary of the results contained in the following chapters.

1.3.1. Chapter 2: The class of kernels.

In Chapter 2 we provide a precise definition of a class of distributions and their associated multipliers. We start with a fixed decomposition $\mathbb{R}^N = \mathbb{R}^{C_1} \oplus \cdots \oplus \mathbb{R}^{C_n}$. Write $\mathbf{x} \in \mathbb{R}^N$ as $(\mathbf{x}_1, \ldots, \mathbf{x}_n)$ with $\mathbf{x}_j = \big(x_k : k \in C_j\big) \in \mathbb{R}^{C_j}$. (Throughout

[2] We will see below in Chapter 7 that if the rank of the matrix $\mathbf{E}$ is greater than 1, then the kernels $\mathcal{K} \in \mathcal{P}(\mathbf{E})$ are integrable at infinity.

[3] A precise definition is given in section 8.1 below.

this paper we use boldface, such as $\mathbf{x}_j$, to indicate a tuple of coordinates, and standard type, such as x_k, to indicate a single coordinate.) Each component $\mathbb{R}^{C_j}$ is equipped with a family of dilations, denoted by $\lambda \cdot \mathbf{x}_j$ for $\lambda > 0$. For simplicity of exposition in this Introduction, we assume that these are the standard isotropic dilations $\lambda \cdot \mathbf{x}_j = \big(\lambda x_k : k \in C_j\big)$ although we will allow more general non-isotropic dilations in Chapter 2 below. If Q_j is the homogeneous dimension of $\mathbb{R}^{C_j}$, then for isotropic dilations, Q_j is the dimension of $\mathbb{R}^{C_j}$.[4] Let n_j be a smooth homogeneous norm on $\mathbb{R}^{C_j}$ so that $n_j(\lambda \cdot \mathbf{x}_j) = \lambda\, n_j(\mathbf{x}_j)$. Thus for isotropic dilations, we can take n_j to be the standard Euclidean norm on $\mathbb{R}^{C_j}$.

The homogeneities on each component $\mathbb{R}^{C_j}$ can be weighted in various ways to give a global family of dilations on $\mathbb{R}^N$. Let $\mathbf{E} = \{e(j,k)\}$ be an $n \times n$ matrix with strictly positive entries, and for $1 \leq j \leq n$, define a family of dilations on $\mathbb{R}^N$ by setting

$$\delta_j(\lambda)[\mathbf{x}] = \big(\lambda^{1/e(j,1)} \cdot \mathbf{x}_1, \ldots, \lambda^{1/e(j,j)} \cdot \mathbf{x}_j, \ldots, \lambda^{1/e(j,n)} \cdot \mathbf{x}_n\big). \tag{1.8}$$

Put

$$N_j(\mathbf{x}) = n_1(\mathbf{x}_1)^{e(j,1)} + \cdots + n_j(\mathbf{x}_j)^{e(j,j)} + \cdots n_n(\mathbf{x}_n)^{e(j,n)} \tag{1.9}$$

so that $N_j\big(\delta_j(\lambda)[\mathbf{x}]\big) = \lambda N_j(\mathbf{x})$. Thus $N_1, \ldots, N_n$ are homogeneous norms[5] on $\mathbb{R}^N$ for the different dilation structures. We require that the components of the matrix $\mathbf{E}$ satisfy

$$\begin{aligned} e(j,j) &= 1 & 1 &\leq j \leq n, \\ e(j,k) &\leq e(j,l)e(l,k) & 1 &\leq j,k,l \leq n. \end{aligned} \tag{1.10}$$

We study a family $\mathcal{P} = \mathcal{P}(\mathbf{E})$ of distributions $\mathcal{K}$ on $\mathbb{R}^N$ which are given away from the origin $0 \in \mathbb{R}^N$ by integration against a smooth function K satisfying the following differential inequalities[6]:

$$\big|\partial_{\mathbf{x}_1}^{\alpha_1} \cdots \partial_{\mathbf{x}_n}^{\alpha_n} K(\mathbf{x})\big| \lesssim \prod_{j=1}^{n} N_j(\mathbf{x})^{-Q_j - |\alpha_j|}. \tag{1.11}$$

(Note that derivatives of K with respect to any of the variables in $\mathbf{x}_j$ are controlled by the family of dilations $\delta_j(\lambda)$ and the homogeneous norm N_j.) In addition, we impose certain cancellation conditions on the distributions $\mathcal{K}$. These are analogous to cancellation conditions for product kernels or flag kernels. Roughly speaking, they require that if $\mathbf{x} = (\mathbf{x}', \mathbf{x}'')$ is a decomposition of the coordinates $\mathbf{x}_1, \ldots, \mathbf{x}_n$ into two subsets, then $\mathcal{K}(\mathbf{x}') = \int \mathcal{K}(\mathbf{x}', \mathbf{x}'')\psi(\mathbf{x}'')\,d\mathbf{x}''$ is a distribution in the $\mathbf{x}''$ variables of the same type. The precise conditions are specified in part (b) of Definition 2.2 below.

Note that in the example of the Heisenberg group given above, we have the natural decomposition $\mathbb{H}^n = \mathbb{C}^n \oplus \mathbb{R} \cong \mathbb{R}^{2n} \oplus \mathbb{R}$. The isotropic dilations on each component are given by $\lambda \cdot (z_1, \ldots, z_n) = (\lambda z_1, \ldots, \lambda z_n)$ and $\lambda \cdot t = \lambda t$, so that the homogeneous dimension of $\mathbb{C}^n$ is $2n$ and the homogeneous dimension of $\mathbb{R}$ is 1.

---

[4] If $\lambda \cdot \mathbf{x}_j = \exp[A_j \log \lambda](\mathbf{x}_j)$ the homogeneous dimension of $\mathbb{R}^{C_j}$ is the trace of A_j.

[5] This is standard terminology, although we only have $N_j(\mathbf{x}+\mathbf{y}) \leq A_j[N_j(\mathbf{x}) + N_j(\mathbf{y})]$ for some constant A_j instead of the usual triangle inequality.

[6] We use standard multi-index notation; see Chapter 2 for more details.

Set $n_1(\mathbf{z}) = |\mathbf{z}| = \left[\sum_{j=1}^n |z_j|^2\right]^{\frac{1}{2}}$ and $n_2(t) = |t|$, so that $n_1(\lambda \cdot \mathbf{z}) = \lambda n_1(\mathbf{z})$ and $n_2(\lambda \cdot t) = \lambda n_2(t)$. If we take $\mathbf{E} = \begin{bmatrix} 1 & 1 \\ 2 & 1 \end{bmatrix}$ then

$$\begin{aligned} N_1(\mathbf{z}, t) &= n_1(\mathbf{z}) + n_2(t) \approx |\mathbf{z}| + |t|, \\ N_2(\mathbf{z}, t) &= n_1(\mathbf{z})^2 + n_2(t) \approx |\mathbf{z}|^2 + |t|, \end{aligned} \tag{1.12}$$

and the corresponding dilations are given by

$$\begin{aligned} \delta_1(\lambda)(\mathbf{z}, t) &= (\lambda \mathbf{z}, \lambda t), \\ \delta_2(\lambda)(\mathbf{z}, t) &= (\lambda^{\frac{1}{2}} \mathbf{z}, \lambda t). \end{aligned}$$

The differential inequalities required by (1.11) are then

$$\left|\partial_{z,\bar{z}}^\alpha \partial_t^\beta K(\mathbf{z}, t)\right| \lesssim N_1(\mathbf{z}, t)^{-2n-|\alpha|} N_2(\mathbf{z}, t)^{-1-\beta} \approx (|\mathbf{z}| + |t|)^{-2n-|\alpha|} (|\mathbf{z}|^2 + |t|)^{-1-\beta}$$

which is in agreement with (1.3).

1.3.2. Chapter 3: Marked partitions.

Our analysis of the distributions $\mathcal{K} \in \mathcal{P}_0(\mathbf{E})$ is based on the decomposition of the unit ball into the regions where one summand in each norm $N_j(\mathbf{x}) = n_1(\mathbf{x}_1)^{e(j,1)} + \cdots + n_n(\mathbf{x}_n)^{e(j,n)}$ is strictly larger than all the others. (There is an analogous decomposition of the Fourier transform $\boldsymbol{\xi}$-space for $\boldsymbol{\xi}$ large, which is needed to understand the differential inequalities of the multiplier $m = \widehat{\mathcal{K}}$.) Matters are simplest in the case when $n = 2$ exemplified by the Heisenberg group, where the two norms are given in equation (1.12). In this case there are three regions near the origin.

1. The set $|\mathbf{z}|^2 \gtrsim |t| \gtrsim |t|^2$, where the isotropic dilations are controlling. Here $N_1(\mathbf{z}, t) \approx |\mathbf{z}| \approx |\mathbf{z}| + |t|$ and $N_2(z, t) \approx |\mathbf{z}|^2 \approx (|\mathbf{z}| + |t|)^2$.
2. The set $|\mathbf{z}| \lesssim |t| \lesssim |t|^{\frac{1}{2}}$, where the automorphic dilations are controlling. Here $N_1(\mathbf{z}, t) \approx |t| \approx |\mathbf{z}|^2 + |t|$ and $N_2(\mathbf{z}, t) \approx |t| \approx |\mathbf{z}|^2 + |t|$.
3. The intermediate region where $|t| \lesssim |\mathbf{z}| \lesssim |t|^{\frac{1}{2}}$ where kernels of the type (1.3) behave like "product kernels". Here $N_1(\mathbf{z}, t) \approx |\mathbf{z}|$ and $N_2(\mathbf{z}, t) \approx |t|$.

The case $n \geq 3$ is more intricate. The required systematic decomposition of the $\mathbf{x}$-space depends on the notion of a 'marked partition' of the set $\{1, \ldots, n\}$, which is a collection of disjoint non-empty subsets $I_1, \ldots, I_s$ of $\{1, \ldots, n\}$ with $\bigcup_{r=1}^s I_r = \{1, \ldots, n\}$, together with a 'marked' element k_r in each subset I_r. This concept arises as follows.

Suppose the matrix $\mathbf{E}$ satisfies the conditions given in equation (1.10), and let $\mathbf{x}$ be a point in the unit ball such that no two terms in any norm $N_j(\mathbf{x})$ are equal. We will see in Proposition 3.3 below that if $n_k(\mathbf{x}_k)$ is the dominant term in $N_j(\mathbf{x})$ for some $j \neq k$ then $n_k(\mathbf{x})$ is also the dominant term in $N_k(\mathbf{x})$. Thus given such a point $\mathbf{x}$, there is a well-defined set $\{k_1, \ldots, k_s\}$ of indices such that $n_{k_r}(\mathbf{x}_{k_r})$ is the dominant term in $N_{k_r}(\mathbf{x})$. For $1 \leq r \leq s$, we let I_r be the set of indices j such that $n_{k_r}(\mathbf{x}_{k_r})$ is dominant in $N_j(\mathbf{x})$. Then $k_r \in I_r$ and this produces a marked partition. The details are given below in Chapter 3.

One may note that the decomposition given via marked partitions is somewhat akin to what happens in the theory of "resolutions of singularities", here applied to (say) a smooth version of a power of the function $F(\mathbf{x}) = \prod_{j=1}^n N_j(\mathbf{x})$. Using that theory (see e.g. [**CGP13**]) one would be led to decompose a neighborhood of the origin into a number of parts, and in each part make an appropriate change of variables so that F becomes essentially a monomial in the new variables. However, in our situation the decomposition by marked partitions is much more explicit and tractable, and no auxiliary change of variables is required.

1.3.3. Chapter 4: Characterization by Fourier transforms.

For $\boldsymbol{\xi} \in \mathbb{R}^N = \mathbb{R}^{C_1} \oplus \cdots \oplus \mathbb{R}^{C_n}$ write $\boldsymbol{\xi} = (\boldsymbol{\xi}_1, \ldots, \boldsymbol{\xi}_n)$ with $\boldsymbol{\xi}_k \in \mathbb{R}^{C_k}$ and set

$$\widehat{N}_j(\boldsymbol{\xi}) = |\boldsymbol{\xi}_1|^{1/e(1,j)} + \cdots + |\boldsymbol{\xi}_j| + \cdots + |\boldsymbol{\xi}_n|^{1/e(n,j)}. \tag{1.13}$$

This is the 'dual norm' to $N_j(\mathbf{x})$. Denote by $\mathcal{M}_\infty(\mathbf{E})$ the class of multipliers $m \in \mathcal{C}^\infty(\mathbb{R}^N)$ satisfying

$$\left|\partial_{\boldsymbol{\xi}_1}^{\boldsymbol{\alpha}_1} \cdots \partial_{\boldsymbol{\xi}_n}^{\boldsymbol{\alpha}_n} m(\boldsymbol{\xi})\right| \lesssim \prod_{j=1}^{n} \left[1 + \widehat{N}_j(\boldsymbol{\xi})\right]^{-|\boldsymbol{\alpha}_j|}. \tag{1.14}$$

On page 4 we introduced the subclass $\mathcal{P}_0(\mathbb{E})$ of proper distributions. In Chapter 4 we show that $\mathcal{K} \in \mathcal{P}_0(\mathbf{E})$ if and only if its Fourier transform $m = \widehat{\mathcal{K}} \in \mathcal{M}_\infty(\mathbf{E})$.

1.3.4. Chapters 5 and 6: Dyadic decompositions.

In Chapter 5 we show that kernels $\mathcal{K} \in \mathcal{P}_0(\mathbf{E})$ can be characterized by their dyadic decomposition into dilates of *bump functions*. Let $\{\varphi^I\}$ be a uniformly bounded family of $\mathcal{C}^\infty$-functions supported in the unit ball, where I runs over the set of all n-tuples $I = (i_1, \ldots, i_n) \in \mathbb{Z}^n$. Let phiI@$[\varphi^I]_I$

$$[\varphi^I]_I(\mathbf{x}) = 2^{-\sum_{j=1}^n Q_j i_j} \varphi^I(2^{-i_1}\mathbf{x}_1, \ldots, 2^{-i_n}\mathbf{x}_n) \tag{1.15}$$

be the L^1-invariant dilation of φ^I, and let

$$\Gamma_{\mathbb{Z}}(\mathbf{E}) = \left\{I = (i_1, \ldots, i_n) \in \mathbb{Z}^n : e(j,k)i_k \leq i_j < 0\right\}. \tag{1.16}$$

If

$$\int_{\mathbb{R}^{C_j}} \varphi^I(\mathbf{x}_1, \ldots, \mathbf{x}_j, \ldots, \mathbf{x}_n)\, d\mathbf{x}_j = 0 \tag{1.17}$$

for each $I \in \Gamma_{\mathbb{Z}}(\mathbf{E})$ and for all $1 \leq j \leq n$, then the infinite series $\sum_{I \in \Gamma_{\mathbb{Z}}(\mathbf{E})} [\varphi^I]_I$ converges in the sense of distributions to a distribution $\mathcal{K} \in \mathcal{P}_0(\mathbf{E})$. More generally, we show that the 'strong' cancellation condition given in (1.17) can be replaced by a variant of the *weak* cancellation condition that was introduced in [**NRSW12**]. Roughly speaking[7], we require that if φ^I does not have cancellation in a variable $\mathbf{x}_j$, then there is a compensating gain in the size of φ^I.

This more general result is then used in our study of convolutions of distributions in Chapter 9. In this study we also need a converse statement: if $\mathcal{K} \in \mathcal{P}_0(\mathbf{E})$, it can be written, modulo distributions corresponding to coarser decompositions of $\mathbb{R}^N$, as a sum $\sum_{I \in \Gamma_{\mathbb{Z}}(\mathbf{E})} [\varphi^I]_I$. This is the subject of Chapter 6.

[7] The precise statement is given in Definition 5.9.

1.3.5. Chapter 7: Significance of the rank of E.

In Chapter 7 we show that the class of classical Calderón-Zygmund kernels corresponds exactly to the case in which the rank of the matrix $\mathbf{E}$ is equal to 1. If the rank is greater than 1, distributions $\mathcal{K} \in \mathcal{P}(\mathbf{E})$ differ from Calderón-Zygmund kernels in two ways:

1) The distribution $\mathcal{K}$ is integrable at infinity, and so we can write $\mathcal{K} = \mathcal{K}_0 + \mathcal{K}_\infty$ with $\mathcal{K}_0, \mathcal{K}_\infty \in \mathcal{P}(\mathbf{E})$, where $\mathcal{K}_0$ has support in the unit ball, and where $\mathcal{K}_\infty$ is given by integration against a function $K_\infty \in \mathcal{C}^\infty(\mathbb{R}^N) \cap L^1(\mathbb{R}^N)$. Thus for $\text{rank}(\mathbf{E}) \geq 2$, we focus on the local behavior of distributions $\mathcal{K} \in \mathcal{P}(\mathbf{E})$. We generally restrict our attention to the class of proper distributions $\mathcal{P}_0(\mathbf{E}) \subset \mathcal{P}(\mathbf{E})$ such that all derivatives are rapidly decreasing at infinity.
2) If $\mathcal{K}$ is a Calderón-Zygmund kernel, then $\big|\{\mathbf{x} \in \mathbb{R}^N : |K(\mathbf{x})| > \lambda\}\big| \lesssim \lambda^{-1}$. If $\mathcal{K} \in \mathcal{P}(\mathbf{E})$ and the rank of $\mathbf{E}$ is greater than 1, such weak type estimates may no longer be true. We show, for example, that if $\text{rank}(\mathbf{E}) = n$ there exist distributions $\mathcal{K} \in \mathcal{P}(\mathbf{E})$ such that $\big|\{\mathbf{x} \in \mathbb{R}^N : |K(\mathbf{x})| > \lambda\}\big| \approx \lambda^{-1}\big[\log \lambda\big]^{n-1}$.
3) In general, there is a notion of 'reduced rank' which determines optimal bounds for estimates of the sort considered in 2). This is discussed in Chapter 7.3.

1.3.6. Chapters 8 and 9: Convolution operators on a Lie group.

In Chapters 8 and 9 we turn to the second part of our analysis dealing with convolution operators $T_\mathcal{K} f = f * \mathcal{K}$ on a homogeneous nilpotent Lie group $G \cong \mathbb{R}^N$. Our two main results are the following:

(A) If $\mathcal{K} \in \mathcal{P}_0(\mathbf{E})$ then each operator $T_\mathcal{K}$ extends to a bounded operator on $L^p(G)$ for $1 < p < \infty$.

(B) If $\mathcal{K}, \mathcal{L} \in \mathcal{P}_0(\mathbf{E})$ then there exists $\mathcal{M} \in \mathcal{P}_0(\mathbf{E})$ such that $T_\mathcal{L} \circ T_\mathcal{K} = T_\mathcal{M}$. Formally $\mathcal{M} = \mathcal{K} * \mathcal{L}$, although the convolution of tempered distributions is not always defined. Thus the space $\mathcal{P}_0(\mathbf{E})$ is an algebra under composition (or convolution).

The first result follows rather easily from the observation that every distribution $\mathcal{K} \in \mathcal{P}_0(\mathbf{E})$ is a flag kernel, since it was established in [**NRSW12**] that convolution with a flag kernel extends to a bounded operator on L^p. The second result is analogous to the result in the same paper that the space of kernels is closed under convolution. However this case is more difficult and requires additional ideas.

We remark that both results are quite simple for Euclidean convolution, since $\widehat{T_\mathcal{K} f}(\boldsymbol{\xi}) = m(\boldsymbol{\xi})\widehat{f}(\boldsymbol{\xi})$ where m is the Fourier transform of $\mathcal{K}$ and hence $m \in \mathcal{M}_\infty(\mathbf{E})$. It is not hard to check that a multiplier $m \in \mathcal{M}_\infty(\mathbf{E})$ satisfies the conditions of the Marcinkiewicz multiplier theorem, and the L^p-boundedness of the operator $T_\mathcal{K}$ follows immediately. Since the product of two such multipliers is again a multiplier of the same class, the result on composition of operators also follows easily in this case.

To prove the results on a more general group, we rely very heavily on the decomposition of distributions $\mathcal{K} \in \mathcal{P}(\mathbf{E})$ as dyadic sums of dilates of bump functions. The heart of the argument is then a careful analysis of sums of the form $\sum_{I,J}[\varphi^I]_I * [\psi^J]_J$.

1.3.7. Chapter 10: Convolution of Calderón-Zygmund kernels.
We study the convolution of Calderón-Zygmund kernels $\mathcal{K}_1, \ldots, \mathcal{K}_m$ with compact support having different homogeneities in Chapter 10. If the homogeneities are suitably adapted to the decomposition on $\mathbb{R}^N$ we show that there is an $s \times s$ matrix $\mathbf{E}$ so that the algebra $\mathcal{P}_0(\mathbf{E})$ contains each $\mathcal{K}_j$. We also establish a kind of converse result: If $\mathbf{E}$ is an $n \times n$ matrix satisfying the basic hypothesis given in (1.10), then there are finitely many different homogeneities on $\mathbb{R}^N$ so that $\mathcal{P}_0(\mathbf{E})$ is the minimal algebra of the above type containing compactly supported Calderón-Zygmund kernels of each homogeneity.

1.3.8. Chapter 11: Regularity of two-flag kernels.
In Chapter 11 we return to the study of kernels belonging to two opposite flags, and show that these belong to $\mathcal{P}_0(\mathbf{E})$ for an appropriate matrix $\mathbf{E}$. In the example studied in section 1.2, where the kernel satisfied the two estimates

$$\left|\partial_x^a \partial_y^b \partial_z^c K(x,y,z)\right| \leq C_{a,b,c} \begin{cases} |x|^{-1-a}(|x|^2+|y|)^{-1-b}(|x|^3+|y|^{3/2}+|z|)^{-1-c} \\ \\ (|x|+|y|^2+|z|^3)^{-1-a}(|y|+|z|^{\frac{3}{2}})^{-1-b}|z|^{-1-c} \end{cases},$$

it turns out that $\mathcal{K} \in \mathcal{P}_0(\mathbf{E})$ where

$$\mathbf{E} = \begin{bmatrix} 1 & 1 & 1 \\ 2 & 1 & 1 \\ 3 & \frac{3}{2} & 1 \end{bmatrix} \quad \text{and} \quad \begin{cases} N_1(x,y,z) &= |x|+|y|+|z|, \\ N_2(x,y,z) &= |x|^2+|y|+|z|, \\ N_3(x,y,z) &= |x|^3+|y|^{3/2}+|z| \end{cases}.$$

1.3.9. Chapters 12 and 13: Operators with variable coefficients.
In Chapters 12 and 13 we describe a wider class of operators with "variable coefficients". An operator T belongs to this extended class if it is of the form

$$T[f](\mathbf{x}) = \int_G K(\mathbf{x},\mathbf{z}) f(\mathbf{x}\cdot\mathbf{z}^{-1})\,d\mathbf{z} \tag{1.18}$$

where for each $\mathbf{x}$, the distribution $K(\mathbf{x},\cdot)$ is a kernel of the type we have been considering that depends in an appropriately smooth way on $\mathbf{x}$. One can then assert:

a) Operators of that type form an algebra, and each is bounded on $L^p(G)$ for $1 < p < \infty$.
b) The standard pseudo-differential operators of order 0 belong to this algebra, and the sub-algebra of these operators is "central" in the sense that an operator of the extended algebra commutes with a standard pseudo-differential operator of order 0, up to an error which is an appropriate "smoothing operator."

CHAPTER 2

The Classes $\mathcal{P}(\mathbf{E})$ and $\mathcal{M}(\mathbf{E})$

In this chapter we define a class of distributions $\mathcal{P}(\mathbf{E})$ and the corresponding class of Fourier multipliers $\mathcal{M}(\mathbf{E})$ on the space $\mathbb{R}^N$ and its dual. We begin by introducing notation that will be used throughout the paper.

2.1. Notation

2.1.1. Decompositions of $\mathbb{R}^N$.

We begin with an initial decomposition of $\mathbb{R}^N$ into direct sums of subspaces where only some of the coordinates $x_1, \dots, x_N$ are non-zero. Thus if $C \subset \{1, \dots, N\}$ is non-empty, set

$$\mathbb{R}^C = \Big\{\mathbf{x} = (x_1, \dots, x_N) \in \mathbb{R}^N : x_j = 0 \text{ for all } j \notin C\Big\} \subset \mathbb{R}^N. \tag{2.1}$$

We will abuse of this notation by allowing $\mathbb{R}^C$ to also indicate the set of c-tuples (c being the cardinality of C) with entries indexed by elements of C. We assume that an initial decomposition of $\mathbb{R}^N$ is given,

$$\mathbb{R}^N = \mathbb{R}^{C_1} \oplus \cdots \oplus \mathbb{R}^{C_n}, \tag{2.2}$$

where $C_1, \dots, C_n \subset \{1, \dots, N\}$ are disjoint non-empty subsets with $\bigcup_{j=1}^n C_j = \{1, \dots, N\}$. Given the decomposition (2.2), if $\mathbf{x} = (x_1, \dots, x_N) \in \mathbb{R}^N$ we write $\mathbf{x} = (\mathbf{x}_1, \dots, \mathbf{x}_n)$ where $\mathbf{x}_j = \{x_l : l \in C_j\}$.

Unfortunately there are further notational complications. We shall sometimes need to consider the direct sum of only a subset of the subspaces $\mathbb{R}^{C_1}, \dots, \mathbb{R}^{C_n}$, and we will also need to partition the set $\{C_1, \dots, C_n\}$ into a union of disjoint subsets and consider the corresponding *coarser* decomposition of $\mathbb{R}^N$. We shall then use the following notation.

(a) If $L = \{l_1, \dots, l_r\} \subset \{1, \dots, n\}$, write

$$\mathbb{R}^L = \mathbb{R}^{C_{l_1}} \oplus \cdots \oplus \mathbb{R}^{C_{l_r}} = \Big\{\mathbf{x} \in \mathbb{R}^N : x_j = 0 \text{ for all } j \notin \bigcup_{k=1}^{r} C_{l_k}\Big\}. \tag{2.3}$$

An element of $\mathbb{R}^L$ can then be written $\mathbf{x}_L = \{\mathbf{x}_k : k \in L\}$.

(b) Let $I_1, \dots, I_s \subset \{1, \dots, n\}$ be disjoint, non-empty subsets with $\bigcup_{r=1}^s I_r = \{1, \dots, n\}$. Write

$$\mathbb{R}^{I_r} = \bigoplus_{k \in I_r} \mathbb{R}^{C_k} = \Big\{\mathbf{x} \in \mathbb{R}^N : x_j = 0 \text{ for all } j \notin \bigcup_{k \in I_r} C_k\Big\}. \tag{2.4}$$

We then have the coarser decomposition $\mathbb{R}^N = \mathbb{R}^{I_1} \oplus \cdots \oplus \mathbb{R}^{I_s}$. If $\mathbf{x} \in \mathbb{R}^N$ we write $\mathbf{x} = (\mathbf{x}_{I_1}, \dots, \mathbf{x}_{I_s})$ where $\mathbf{x}_{I_r} \in \mathbb{R}^{I_r}$.

Note that we have slightly abused notation since the symbols $\mathbb{R}^L$ and $\mathbb{R}^{I_r}$ in equations (2.3) and (2.4) have a different meaning than the symbol $\mathbb{R}^C$ in equation (2.1). However C is a subset of $\{1,\dots,N\}$ while L and I_r are subsets of $\{1,\dots,n\}$.

2.1.2. Dilations on $\mathbb{R}^N$.
The underlying space $\mathbb{R}^N$ is equipped with a family of dilations given by

$$\lambda\cdot\mathbf{x}=\lambda\cdot(x_1,\dots,x_N)=\left(\lambda^{d_1}x_1,\dots,\lambda^{d_N}x_N\right) \tag{2.5}$$

where each $d_j>0$. We can restrict these dilations in equation (2.5) to each subspace $\mathbb{R}^{C_j}$ and write $\lambda\cdot\mathbf{x}_j=\left\{\lambda^{d_l}x_l : l\in C_j\right\}$. The homogeneous dimension of the subspace $\mathbb{R}^{C_j}$ is then $Q_j=\sum_{l\in C_j}d_l$. Let $n_j:\mathbb{R}^{C_j}\to[0,\infty)$ be a smooth homogeneous norm for this family of dilations so that

$$n_j(\mathbf{x}_j)\approx\sum_{l\in C_j}|x_l|^{1/d_l}. \tag{2.6}$$

Thus $n_j(\lambda\cdot\mathbf{x}_j)=\lambda n_j(\mathbf{x}_j)$ for $\lambda>0$. Put

$$\mathbb{B}(\rho)=\left\{\mathbf{x}\in\mathbb{R}^N : n_j(\mathbf{x}_j)<\rho,\ 1\le j\le n\right\}.$$

We use standard multi-index notation for derivatives. If $\boldsymbol{\gamma}=(\gamma_1,\dots,\gamma_N)\in\mathbb{N}^N$ then $\partial_{\mathbf{x}}^{\boldsymbol{\gamma}}=\prod_{l=1}^N\partial_{x_l}^{\gamma_l}$ and $|\gamma|=\sum_{l=1}^N\gamma_l$. Using the decomposition (2.2), we can also write $\boldsymbol{\gamma}=(\boldsymbol{\gamma}_1,\dots,\boldsymbol{\gamma}_n)\in\mathbb{N}^{C_1}\times\cdots\times\mathbb{N}^{C_n}$. When estimating derivatives $\partial_{\mathbf{x}}^{\boldsymbol{\gamma}}f(\mathbf{x})$ in terms of homogeneous norms on each subspace $\mathbb{R}^{C_j}$ we need to use an appropriately weighted length. Thus put

$$[\![\boldsymbol{\gamma}_j]\!]=\sum_{l\in C_j}\gamma_l d_l \qquad\text{and}\qquad [\![\boldsymbol{\gamma}]\!]=\sum_{j=1}^n[\![\boldsymbol{\gamma}_j]\!]=\sum_{l=1}^N\gamma_l d_l. \tag{2.7}$$

The space of infinitely differentiable real-valued functions on $\mathbb{R}^N$ with compact support is denoted by $\mathcal{C}_0^\infty(\mathbb{R}^N)$ and the space of Schwartz functions is denoted by $\mathcal{S}(\mathbb{R}^N)$. The basic semi-norms on these spaces are given by

$$\begin{aligned}
||\varphi||_{(m)} &= \sup_{\mathbf{x}\in\mathbb{R}^N}\left\{\left|\partial_{\mathbf{x}}^{\boldsymbol{\gamma}}\varphi(\mathbf{x})\right| : |\gamma|\le m\right\} && \text{for } \varphi\in\mathcal{C}_0^\infty(\mathbb{R}^N),\\
||\psi||_M &= \sup_{\mathbf{x}\in\mathbb{R}^N}\left\{\left|(1+|\mathbf{x}|)^{\boldsymbol{\alpha}}\partial_{\mathbf{x}}^{\boldsymbol{\beta}}\psi(\mathbf{x})\right| : |\boldsymbol{\alpha}|+|\boldsymbol{\beta}|\le M\right\} && \text{for } \psi\in\mathcal{S}(\mathbb{R}^N).
\end{aligned}$$

Let $\mathcal{A}$ be an index set. We use the following terminology from Chapter 4 in [**NRSW12**].

1) If $\{\varphi_\alpha : \alpha\in\mathcal{A}\}\subset\mathcal{C}_0^\infty(\mathbb{R}^N)$, then the family $\{\varphi_\alpha\}$ is *normalized* in terms of a function $\Phi\in\mathcal{C}_0^\infty(\mathbb{R}^N)$ if there are constants $C, C_m>0$ and integers p_m with the following properties.
   a) If the support of Φ is contained in the ball $\mathbb{B}(\rho)$ then the support of each φ_α is contained in the ball $\mathbb{B}(C\rho)$.
   b) For every $m\in\mathbb{N}$ and every $\alpha\in\mathcal{A}$, $||\varphi_\alpha||_{(m)}\le C_m||\Phi||_{(m+p_m)}$.
2) A family $\{\varphi_\alpha : \alpha\in\mathcal{A}\}\subset\mathcal{C}_0^\infty(\mathbb{R}^N)$ is *uniformly bounded* if all the members are supported in a fixed ball $\mathbb{B}(\rho)$ and if there are constants $C_m>0$ so that $||\varphi_\alpha||_{(m)}\le C_m$ for all $\alpha\in\mathcal{A}$.

3) If $\{\psi_\alpha : \alpha \in \mathcal{A}\} \subset \mathcal{S}(\mathbb{R}^N)$, then the family $\{\psi_\alpha\}$ is *normalized* in terms of a function $\Psi \in \mathcal{S}(\mathbb{R}^N)$ if there are constants $C_M > 0$ and integers $p_M \in \mathbb{N}$ so that $||\psi_\alpha||_{[M]} \leq C_M ||\Psi||_{[M+p_M]}$ for every $M \in \mathbb{N}$.
4) A family $\{\psi_\alpha : \alpha \in \mathcal{A}\} \subset \mathcal{S}(\mathbb{R}^N)$ is *uniformly bounded* if there are constants $C_M > 0$ so that $||\psi_\alpha||_M \leq C_M$ for all $\alpha \in \mathcal{A}$.

Given the decomposition $\mathbb{R}^N = \mathbb{R}^{C_1} \oplus \cdots \oplus \mathbb{R}^{C_n}$, define an n-parameter family of dilations by setting $(\lambda_1, \ldots, \lambda_n) \cdot \mathbf{x} = (\lambda_1 \cdot \mathbf{x}_1, \ldots, \lambda_n \cdot \mathbf{x}_n)$. If $(\lambda_1, \ldots, \lambda_n) = (2^{-t_1}, \ldots, 2^{-t_n})$ with $\mathbf{t} = (t_1, \ldots, t_n) \in \mathbb{R}^n$ then for $f \in L^1(\mathbb{R}^N)$, we define

$$(2.8)\qquad \begin{aligned} (f)_{\mathbf{t}}(\mathbf{x}) &= f(2^{-t_1} \cdot \mathbf{x}_1, \ldots, 2^{-t_n} \cdot \mathbf{x}_n), \\ [f]_{\mathbf{t}}(\mathbf{x}) &= 2^{-\sum_{j=1}^n t_j Q_j} f(2^{-t_1} \cdot \mathbf{x}_1, \ldots, 2^{-t_n} \cdot \mathbf{x}_n). \end{aligned}$$

Note that $||(f)_{\mathbf{t}}||_{L^\infty(\mathbb{R}^N)} = ||f||_{L^\infty(\mathbb{R}^N)}$ while $||[f]_{\mathbf{t}}||_{L^1(\mathbb{R}^N)} = ||f||_{L^1(\mathbb{R}^N)}$.

2.2. Global norms

The classes $\mathcal{P}(\mathbf{E})$ and $\mathcal{M}(\mathbf{E})$ of distributions and multipliers are defined using families of norms $\{N_1, \ldots, N_n\}$ and dual norms $\{\widehat{N}_1, \ldots, \widehat{N}_n\}$. These in turn are defined in terms of an $n \times n$ matrix $\mathbf{E} = \{e(j,k)\}$, where each entry $e(j,k) \in (0, \infty)$. For $1 \leq j \leq n$ set

$$(2.9)\qquad \begin{aligned} N_j(\mathbf{t}_1, \ldots, \mathbf{t}_n) &= n_1(\mathbf{t}_1)^{e(j,1)} + \cdots + n_n(\mathbf{t}_n)^{e(j,n)} \approx \sum_{k=1}^n \sum_{l \in C_k} |t_l|^{e(j,k)/d_l}, \\ \widehat{N}_j(\mathbf{t}_1, \ldots, \mathbf{t}_n) &= n_1(\mathbf{t}_1)^{\frac{1}{e(1,j)}} + \cdots + n_n(\mathbf{t}_n)^{\frac{1}{e(n,j)}} \approx \sum_{k=1}^n \sum_{l \in C_k} |t_l|^{1/e(k,j) d_l}. \end{aligned}$$

The entries of $\mathbf{E}$ are usually subject to certain natural constraints.[1]

DEFINITION 2.1. The matrix $\mathbf{E}$ satisfies the basic hypothesis if :

$$(2.10)\qquad \begin{aligned} e(j,j) &= 1, && \text{for } 1 \leq j \leq n, \\ e(j,l) &\leq e(j,k)e(k,l) && \text{for } 1 \leq j,k,l \leq n. \end{aligned}$$

In particular, $1 \leq e(j,k)e(k,j)$ for $1 \leq j,k \leq n$.

We will need to restrict the norms $\{N_j\}$ to certain subspaces of $\mathbb{R}^N$. Thus if $L = \{l_1, \ldots, l_s\} \subset \{1, \ldots, n\}$, we use the notation introduced on page 11 and set $\mathbb{R}^L = \bigoplus_{l \in L} \mathbb{R}^{C_l}$. For $\mathbf{t}_L = (\mathbf{t}_{l_1}, \ldots, \mathbf{t}_{l_s}) \in \mathbb{R}^L$, and for each $l \in L$, put

$$N_l(\mathbf{t}_L) = \sum_{m \in L} n_m(\mathbf{t}_m)^{e(l,m)}.$$

If $R = \{R_1, \ldots, R_s\}$ are positive real numbers and if $\psi \in \mathcal{C}_0^\infty(\mathbb{R}^L)$, define $\psi_R \in \mathcal{C}_0^\infty(\mathbb{R}^L)$ by setting

$$\psi_R(\mathbf{t}_L) = \psi(R_1 \cdot \mathbf{t}_{l_1}, \ldots, R_s \cdot \mathbf{t}_{l_s}).$$

[1]The motivation for these constraints is discussed in Appendix 14 below.

2.3. Classes of distributions and multipliers

Let $\mathbb{R}^N = \mathbb{R}^{C_1} \oplus \cdots \oplus \mathbb{R}^{C_n}$, let $\mathbf{E} = \{e(j,k)\}$ be an $n \times n$ matrix satisfying (2.10), and let $\{N_1, \ldots, N_n\}$ and $\{\widehat{N}_1, \ldots, \widehat{N}_n\}$ be the norms and dual defined in (2.9).

DEFINITION 2.2. A distribution $\mathcal{K}$ on $\mathbb{R}^N$ belongs to the class $\mathcal{P}(\mathbf{E})$ if it satisfies the following differential inequalities and cancellation conditions.

(a) [Differential Inequalities] Away from the origin the distribution $\mathcal{K}$ is given by integration against a $\mathcal{C}^\infty$-function K, and for every $\boldsymbol{\gamma} = (\gamma_1, \ldots, \gamma_n) \in \mathbb{N}^{C_1} \times \cdots \times \mathbb{N}^{C_n}$, there is a constant $C_{\boldsymbol{\gamma}}$ so that

$$|\partial^{\boldsymbol{\gamma}} K(\mathbf{x})| \leq C_{\boldsymbol{\gamma}} \prod_{j=1}^{n} N_j(\mathbf{x})^{-(Q_j + [\![\gamma_j]\!])}.$$

(b) [Cancellation Conditions] Let $L = \{l_1, \ldots, l_r\}$ and $M = \{m_1, \ldots, m_s\}$ be any pair of complementary subsets of $\{1, \ldots, n\}$, let $R = \{R_1, \ldots, R_s\}$ be any positive real numbers, and let $\psi \in \mathcal{C}_0^\infty(\mathbb{R}^M)$ have support in the unit ball. Define a distribution $\mathcal{K}_{\psi,R}$ on $\mathbb{R}^L$ by setting

$$\langle \mathcal{K}_{\psi,R}, \varphi \rangle = \langle \mathcal{K}, \varphi \otimes \psi_R \rangle$$

for every $\varphi \in \mathcal{C}_0^\infty(\mathbb{R}^L)$. Then $\mathcal{K}_{\psi,R}$ uniformly satisfies the analogue of the estimates in (a) on the space $\mathbb{R}^L$: precisely, this means that away from the origin of $\mathbb{R}^L$ the distribution $\mathcal{K}_{\psi,R}$ is given by integration against a smooth function $K_{\psi,R}$ and for every $\boldsymbol{\gamma} = (\gamma_{l_1}, \ldots, \gamma_{l_r}) \in \mathbb{N}^{C_{l_1}} \times \cdots \times \mathbb{N}^{C_{l_r}}$ there is a constant $C'_{\boldsymbol{\gamma}}$ depending only on the constants $\{C_{\boldsymbol{\gamma}}\}$ in (a) and the norms $\{||\psi||_{(m)}\}$ (but independent of R) so that

$$|\partial^{\boldsymbol{\gamma}} K_{\psi,R}(\mathbf{x})| \leq C'_{\boldsymbol{\gamma}} \prod_{t=1}^{r} N_{l_t}(\mathbf{x}_L)^{-(Q_{l_t} + [\![\gamma_{l_t}]\!])}.$$

In particular, if $L = \emptyset$ then $|\langle \mathcal{K}, \psi_R \rangle|$ is bounded independently of R.

REMARK 2.3. Since $N_j(\mathbf{x})^{-[\![\gamma_j]\!]} \leq n_j(\mathbf{x}_j)^{-[\![\gamma_j]\!]}$, it follows that a distribution $\mathcal{K} \in \mathcal{P}(\mathbf{E})$ is a product kernel on $\mathbb{R}^N = \bigoplus_{j=1}^n \mathbb{R}^{C_j}$ in the sense of [**NRS01**].[2] (The cancellation conditions follow from the Fourier transform characterization. See [**NRS01**], Remark 2.16.) However, the distributions in $\mathcal{P}(\mathbf{E})$ are in general more regular; they are singular only at the origin.

DEFINITION 2.4. If $m \in \mathcal{C}^\infty(\mathbb{R}^N \setminus \{0\})$ then m belongs to the class of multipliers $\mathcal{M}(\mathbf{E})$ if for every $\boldsymbol{\gamma} = (\gamma_1, \ldots, \gamma_n) \in \mathbb{N}^{C_1} \times \cdots \times \mathbb{N}^{C_n}$ there is a constant $C_{\boldsymbol{\gamma}}$ so that

$$|\partial^{\boldsymbol{\gamma}} m(\boldsymbol{\xi})| \leq C_{\boldsymbol{\gamma}} \prod_{j=1}^{n} \widehat{N}_j(\boldsymbol{\xi})^{-[\![\gamma_j]\!]}.$$

We shall see in Lemma 7.3 that unless $e(j,k) = e(j,l)e(l,k)$ for all $j,k,l \in \{1, \ldots, n\}$, the distribution $\mathcal{K} \in \mathcal{P}(\mathbf{E})$ is integrable at infinity. We are primarily interested in the local behavior of these kernels and the following restricted classes of distributions and multipliers.

---

[2] In the paper [**NRS01**] the homogeneous norm $n_j(\mathbf{x}_j)$ is denoted simply by $|\mathbf{x}_j|$.

DEFINITION 2.5. Let $|\mathbf{x}|$ denote the usual Euclidean norm of a vector $\mathbf{x} \in \mathbb{R}^N$.

(a) A distribution $\mathcal{K}$ on $\mathbb{R}^N$ belongs to the class $\mathcal{P}_0(\mathbf{E})$ if it belongs to the class $\mathcal{P}(\mathbf{E})$ and away from the origin it is given by a smooth function K satisfying the differential inequalities

$$\big|\partial_{\mathbf{x}}^{\gamma} K(\mathbf{x})\big| \leq C_{M,\gamma} \prod_{j=1}^{n} N_j(\mathbf{x})^{-(Q_j+[\![\gamma_j]\!])} \big(1+|\mathbf{x}|\big)^{-M} \tag{2.11}$$

for every multi-index $\gamma = (\gamma_1, \ldots, \gamma_n) \in \mathbb{N}^{C_1} \times \cdots \times \mathbb{N}^{C_n}$ and every $M \geq 0$. Moreover the quotient distributions $\mathcal{K}_{\psi,R}$ defined in part (b) of Definition 2.2 satisfy

$$\big|\partial^{\gamma} K_{\psi,R}(\mathbf{x})\big| \leq D_{M,\gamma} \big(1+|\mathbf{x}|\big)^{-M} \prod_{t=1}^{r} N_{l_t}^{L}(\mathbf{x}_L)^{-(Q_{l_t}+[\![\gamma_{l_t}]\!])}. \tag{2.12}$$

(b) A function $m \in \mathcal{C}^{\infty}(\mathbb{R}^N)$ belongs to $\mathcal{M}_{\infty}(\mathbf{E})$ if there is a constant C_γ so that

$$\Big|\partial_{\boldsymbol{\xi}}^{\gamma} m(\boldsymbol{\xi})\Big| \leq C_\gamma \prod_{j=1}^{n} \big(1+\widehat{N}_j(\boldsymbol{\xi})\big)^{-[\![\gamma_j]\!]}$$

for every multi-index $\gamma = (\gamma_1, \ldots, \gamma_n) \in \mathbb{N}^{C_1} \times \cdots \times \mathbb{N}^{C_n}$.

We will see in Chapter 4 below that the multipliers $m \in \mathcal{M}_{\infty}(\mathbf{E})$ are precisely the Fourier transforms of the distributions $\mathcal{K} \in \mathcal{P}_0(\mathbf{E})$, and conversely, that distributions $\mathcal{K} \in \mathcal{P}_0(\mathbf{E})$ are the inverse Fourier transforms of multipliers $m \in \mathcal{M}_{\infty}(\mathbf{E})$. The family of norms

$$|\mathcal{K}|_M = \sum_{|\gamma|,|\gamma'|\leq M} \inf\Big\{C_{M,\gamma} + D_{M,\gamma'} : \text{ (2.11) and (2.12) hold}\Big\} \tag{2.13}$$

with $M \in \mathbb{N}$ induces a Fréchet space topology on the space $\mathcal{P}_0(\mathbf{E})$.

CHAPTER 3

Marked partitions and decompositions of $\mathbb{R}^N$

3.1. Dominant terms in $N_j(\mathbf{t})$ and $\widehat{N}_j(\mathbf{t})$

To study the kernels and multipliers introduced in Chapter 2, we need to partition the space $\mathbb{R}^N$ into regions where the norms $N_j(\mathbf{t}) = \sum_{k=1}^n n_k(\mathbf{t}_k)^{e(j,k)}$ or $\widehat{N}_j(\mathbf{t}) = \sum_{k=1}^n n_k(\mathbf{t}_k)^{1/e(k,j)}$ are comparable to a single term $n_k(\mathbf{t}_k)^{e(j,k)}$ or $n_k(\mathbf{t}_k)^{1/e(k,j)}$. We begin by introducing the notion of strict dominance and A-dominance.

DEFINITION 3.1. Let $A \geq 1$.

(a) Let $\mathbf{t} \in \mathbb{B}(1)$. The term $\mathbf{t}_k$ is *A-dominant in* $N_j(\mathbf{t})$ if for every $l \neq k$

$$n_l(\mathbf{t}_l)^{e(j,l)} < \big[A\, n_k(\mathbf{t}_k)\big]^{e(j,k)}.$$

The term $\mathbf{t}_k$ is *strictly dominant in* $N_j(\mathbf{t})$ if we can take $A = 1$.

(b) Let $\mathbf{t} \in \mathbb{B}(1)^c$. The term $\mathbf{t}_k$ is *A-dominant* in $\widehat{N}_j(\mathbf{t})$ if for every $l \neq k$

$$n_l(\mathbf{t}_l)^{1/e(l,j)} < \big[A\, n_k(\mathbf{t}_k)\big]^{1/e(k,j)}.$$

The term $\mathbf{t}_k$ is strictly dominant in $\widehat{N}_j(\mathbf{t})$ if we can take $A = 1$.

REMARKS 3.2.

1) If $\mathbf{t} \in \mathbb{B}(1)$ and if $\mathbf{t}_k$ is A-dominant in $N_j(\mathbf{t})$ then $N_j(\mathbf{t}) \approx n_k(\mathbf{t}_k)^{e(j,k)}$ where the implied constant depends only on A, $\mathbf{E}$, and n.
2) If $\mathbf{t} \in \mathbb{B}(1)^c$ and if $\mathbf{t}_k$ is A-dominant in $\widehat{N}_j(\mathbf{t})$ then $\widehat{N}_j(\mathbf{t}) \approx n_k(\mathbf{t})^{1/e(k,j)}$ where the implied constant depends only on A, $\mathbf{E}$, and n.
3) If $\mathbf{t} \in \mathbb{B}(1)^c$ then each $\widehat{N}_j(\mathbf{t}) \geq 1$.
4) If $\mathbf{t} \in \mathbb{B}(1)^c$ and if $\mathbf{t}_k$ is A-dominant in any $\widehat{N}_j(\mathbf{t})$ then $An_k(\mathbf{t}_k) > 1$.

The following is a simple but key fact about dominance.

PROPOSITION 3.3. Suppose the matrix $\mathbf{E}$ satisfies the basic hypothesis (2.10).

(a) Let $\mathbf{t} \in \mathbb{B}(1)$. If $\mathbf{t}_k$ is A-dominant in $N_j(\mathbf{t})$ then $\mathbf{t}_k$ is also A-dominant in $N_k(\mathbf{t})$. In particular, if $\mathbf{t}_k$ is strictly dominant in $N_j(\mathbf{t})$, then $\mathbf{t}_k$ is strictly dominant in $N_k(\mathbf{t})$.

(b) Let $\mathbf{t} \in \mathbb{B}(1)^c$. If $\mathbf{t}_k$ is A-dominant in $\widehat{N}_j(\mathbf{t})$ then $\mathbf{t}_k$ is also A-dominant in $\widehat{N}_k(\mathbf{t})$. In particular if $\mathbf{t}_k$ is strictly dominant in $\widehat{N}_j(\mathbf{t})$ then $\mathbf{t}_k$ is also strictly dominant in $\widehat{N}_k(\mathbf{t})$.

PROOF. Let $\mathbf{t} \in \mathbb{B}(1)$ and $A \geq 1$. Then $e(k,l) \geq e(j,l)/e(j,k)$ by the basic hypothesis, and since $n_l(\mathbf{t}_l) \leq 1$ it follows that $n_l(\mathbf{t}_l)^{e(k,l)} \leq n_l(\mathbf{t}_l)^{e(j,l)/e(j,k)}$. If $\mathbf{t}_k$

is A-dominant in $N_j(\mathbf{t})$ and if $l \neq k$, then $n_l(\mathbf{t}_l)^{e(j,l)} < \big[A\, n_k(\mathbf{t}_k)\big]^{e(j,k)}$, and so

$$n_l(\mathbf{t}_l)^{e(k,l)} \leq n_l(\mathbf{t}_l)^{e(j,l)/e(j,k)} < A\, n_k(\mathbf{t}_k).$$

Thus $\mathbf{t}_k$ is A-dominant in $N_k(\mathbf{t})$. On the other hand, if $\mathbf{t} \in \mathbb{B}(1)^c$ and $\mathbf{t}_k$ is A-dominant in $\widehat{N}_j(\mathbf{t})$, then every $l \neq k$ we have $n_l(\mathbf{t}_l)^{1/e(l,j)} < [An_k(\mathbf{t}_k)]^{1/e(k,j)}$. Since $e(l,j) \leq e(l,k)e(k,j)$ and $An_k(\mathbf{t}_k) > 1$ we have

$$n_l(\mathbf{t}_l)^{1/e(l,k)} = \Big[n_l(\mathbf{t}_l)^{1/e(l,j)}\Big]^{e(l,j)/e(l,k)} < \Big[An_k(t_k)\Big]^{e(l,j)/e(l,k)e(k,j)} \leq A\, n_k(\mathbf{t}_k).$$

Thus $\mathbf{t}_k$ is A-dominant in $\widehat{N}_k(\mathbf{t})$. □

3.2. Marked partitions and the sets E_S and $\widehat{E}_S$

The notion of strict dominance allows us to decompose (up to a closed set of measure zero) the ball $\mathbb{B}(1) = \big\{\mathbf{t} \in \mathbb{R}^N : n_j(\mathbf{t}_j) \leq 1\big\}$ into sets where one term in each N_j is strictly dominant. The same is true for the complement $\mathbb{B}(1)^c$ and the norms $\widehat{N}_j$. The decomposition depends on the notion of a *marked partition.*

DEFINITION 3.4.

(a) A *marked partition* $S = \big\{(I_1, k_1); \ldots ; (I_s; k_s)\big\}$ of the set $\{1, \ldots, n\}$ is a collection $\{I_1, \ldots, I_s\}$ of disjoint non-empty subsets of $\{1, \ldots, n\}$ such that $\bigcup_{r=1}^{s} I_r = \{1, \ldots, n\}$, together with a choice of a particular element $k_r \in I_r$ in each subset.

(b) The partition $S_0 = \big\{(\{1\}, 1); \ldots ; (\{n\}, n)\big\}$ is called the *principal partition* of $\{1, \ldots, n\}$.

(c) Any marked partition $S = \big\{(I_1, k_1); \ldots ; (I_s; k_s)\big\}$ with $s < n$ is called a *non-principal partition.*

(d) The set of all marked partitions of $\{1, \ldots, n\}$ is denoted by $\mathcal{S}(n)$.

We want to associate a unique marked partition to every point $\mathbf{t} \in \mathbb{R}^N$, but to do this we need to exclude the closed set of measure zero where two different terms in $N_j(\mathbf{t})$ or $\widehat{N}_j(\mathbf{t})$ can be equal. Let

$$\Xi = \Big\{\mathbf{t} \in \mathbb{R}^N : \text{there exists } j \text{ and } l \neq m \text{ so that}$$
$$n_l(\mathbf{t}_l)^{e(j,l)} = n_m(\mathbf{t}_m)^{e(j,m)} \text{ or } n_l(\mathbf{t}_l)^{1/e(l,j)} = n_m(\mathbf{t}_m)^{1/e(m,j)}\Big\}.$$

The set Ξ is closed and has measure zero, and $\mathbb{R}^N \setminus \Xi$ is dense in $\mathbb{R}^N$. Note that if $t \notin \Xi$, all the summands in $N_j(\mathbf{t})$ and $\widehat{N}_j(\mathbf{t})$ are distinct.

We associate to every point $\mathbf{t} \in \mathbb{R}^N \setminus \Xi$ a marked partition $S(\mathbf{t}) \in \mathcal{S}(n)$. There are two cases, depending on whether the point $\mathbf{t}$ belongs to $\mathbb{B}(1)$ or to $\mathbb{B}(1)^c$.

Case 1: If $\mathbf{t} \in \mathbb{B}(1) \setminus \Xi$ then for every j the terms $n_1(\mathbf{t}_1)^{e(j,1)}, \ldots, n_n(\mathbf{t}_n)^{e(j,n)}$ in $N_j(\mathbf{t})$ are distinct. Thus for each $j \in \{1, \ldots, n\}$ there exists a unique integer $k(j) \in \{1, \ldots, n\}$ such that $\mathbf{t}_{k(j)}$ is strictly dominant in $N_j(\mathbf{t})$. Let $\{k_1, \ldots, k_s\}$ be the set of distinct integers $k(j)$ which arise in this way. For $1 \leq r \leq s$ set

$$I_r = \Big\{j \in \{1, \ldots, n\} : \mathbf{t}_{k_r} \text{ is strictly dominant in } N_j(\mathbf{t})\Big\}.$$

Every j belongs to a unique I_r since there is only one term in $N_j(\mathbf{t})$ which is dominant, and according to Proposition 3.3, we have $k_r \in I_r$. Thus $S(\mathbf{t}) = \big\{(I_1, k_1); \ldots ; (I_s, k_s)\big\} \in \mathcal{S}(n)$.

Case 2: If $\mathbf{t} \in \mathbb{B}(1)^c \setminus \Xi$, the terms $n_1(\mathbf{t}_1)^{1/(1,j)}, \ldots, n_n(\mathbf{t}_n)^{1/e(n,j)}$ in $\widehat{N}_j(\mathbf{t})$ are distinct, and again for each j there exists a unique integer $k(j) \in \{1, \ldots, n\}$ such that $\mathbf{t}_{k(j)}$ is strictly dominant in $\widehat{N}_j(\mathbf{t})$. Let $\{k_1, \ldots, k_s\}$ be the set of distinct integers that arise, and again set

$$I_r = \Big\{ j \in \{1, \ldots, n\} : \mathbf{t}_{k_r} \text{ is strictly dominant in } \widehat{N}_j(\mathbf{t}) \Big\}.$$

We have $k_r \in I_r$ by Proposition 3.3, and thus $S = S(\mathbf{t}) = \{(I_1, k_1); \ldots; (I_s, k_s)\} \in \mathcal{S}(n)$.

DEFINITION 3.5. Let $S = \big\{(I_1, k_1); \ldots; (I_s, k_s)\big\} \in \mathcal{S}(n)$, and let $A \geq 1$. Then

$$\begin{aligned} E_S^A &= \Big\{ \mathbf{t} = (\mathbf{t}_1, \ldots, \mathbf{t}_n) \in \mathbb{B}(1) : \text{ for all } r \in \{1, \ldots, s\} \text{ and } j \in I_r, \\ &\qquad\qquad \mathbf{t}_{k_r} \text{ is } A\text{-dominant in } N_j(\mathbf{t})) \Big\} \\ &= \Big\{ \mathbf{t} \in \mathbb{B}(1) : \text{for all } r \in \{1, \ldots, s\}, \text{ all } j \in I_r, \text{ and all } l \neq k_r, \\ &\qquad\qquad n_l(\mathbf{t}_l)^{e(j,l)} < \big[A\, n_{k_r}(\mathbf{t}_{k_r})\big]^{e(j,k_r)} \Big\}; \end{aligned}$$

$$\begin{aligned} \widehat{E}_S^A &= \Big\{ \mathbf{t} = (\mathbf{t}_1, \ldots, \mathbf{t}_n) \in \mathbb{B}(1)^c : \text{ for all } r \in \{1, \ldots, s\} \text{ and } j \in I_r, \\ &\qquad\qquad \mathbf{t}_{k_r} \text{ is } A\text{-dominant in } \widehat{N}_j(\mathbf{t})) \Big\} \\ &= \Big\{ \mathbf{t} \in \mathbb{B}(1)^c : \text{for all } r \in \{1, \ldots, s\}, \text{ all } j \in I_r, \text{ and all } l \neq k_r, \\ &\qquad\qquad n_l(\mathbf{t}_l)^{1/e(l,j)} < \big[A\, n_{k_r}(\mathbf{t}_{k_r})\big]^{1/e(k_r,j)} \Big\}. \end{aligned}$$

When $A = 1$ we write $E_S^1 = E_S$ and $\widehat{E}_S^1 = \widehat{E}_S$.

PROPOSITION 3.6. The set E_S^A and $\widehat{E}_S^A$ are open. Moreover,

$$\mathbb{B}(1) \setminus \Xi = \bigcup_{S \in \mathcal{S}(n)} E_S, \qquad \mathbb{B}(1)^c \setminus \Xi = \bigcup_{S \in \mathcal{S}(n)} \widehat{E}_S.$$

If $A > 1$ then

$$\mathbb{B}(1) \subset \bigcup_{S \in \mathcal{S}(n)} E_S^A, \qquad \mathbb{B}(1)^c \subset \bigcup_{S \in \mathcal{S}(n)} \widehat{E}_S^A.$$

PROOF. The first line follows from the construction of E_S and $\widehat{E}_S$. The second line follows since the closure of $\mathbb{R}^N \setminus \Xi$ is all of $\mathbb{R}^N$, and if $A > 1$ then E_S^A is an open neighborhood of the closure of E_S, and $\widehat{E}_S^A$ is an open neighborhood of the closure of $\widehat{E}_S$. □

3.3. Characterizing the sets E_S^A and $\widehat{E}_S^A$

Let $S = \big\{(I_1, k_1); \ldots; (I_s, k_s)\big\} \in \mathcal{S}(n)$. The definitions of the sets E_S^A and $\widehat{E}_S^A$ involve inequalities between the norms $n_j(\mathbf{t}_j)$ and $n_{k_r}(\mathbf{t}_{k_r})$ for all $r \in \{1, \ldots, s\}$ and *all* $j \neq k_r$. In this section we show that the sets are characterized by a smaller number of inequalities relating, for given r, $n_{k_r}(\mathbf{t}_{k_r})$ to $n_j(\mathbf{t}_j)$ *only* for $j \in I_r \cup \{k_1, \ldots, k_s\}$.

DEFINITION 3.7. Let $\{(I_1,k_1);\dots;(I_s,k_s)\}\in\mathcal{S}(n)$. For $1\le r\le s$ and $1\le l\le n$ set

$$\sigma_S(k_r,l)=\min_{j\in I_r}\frac{e(j,l)}{e(j,k_r)},\qquad \tau_S(l,k_r)=\min_{j\in I_r}\frac{e(l,j)}{e(k_r,j)}.$$

PROPOSITION 3.8. Suppose that the matrix $\mathbf{E}=\{e(j,k)\}$ satisfies the basic hypothesis (2.10) and that $S=\big((I_1,k_1);\dots;(I_s,k_s)\big)\in\mathcal{S}(n)$.

(a) If $1\le r\le s$ and if $l\in I_r$ then

$$\sigma_S(k_r,l)=e(l,k_r)^{-1},\qquad \tau_S(l,k_r)=e(k_r,l)^{-1}.$$

(b) Let $1\le r,p\le s$. If $m\in I_p$ then

$$\sigma_S(k_r,m)\ge\sigma_S(k_r,k_p)\sigma_S(k_p,m),\qquad \tau_S(m,k_r)\ge\tau_S(m,k_p)\tau_S(k_p,k_r).$$

PROOF. It suffices to prove the statements about σ_S; the statements about τ_S follow in the same way. If $l\in I_r$ we can take $j=l$ in the definition of $\sigma_S(k_r,l)$ to see that $\sigma_S(k_r,l)\le e(l,k_r)^{-1}$. On the other hand it follows from the basic assumption (2.10) that $e(j,l)/e(j,k_r)\ge e(l,k_r)^{-1}$ for any l, and this gives the opposite inequality and establishes the first equality in (a).

Next let $m\in I_p$. Then using (a) and the basic hypothesis, if $j_0\in I_r$

$$\begin{aligned}\sigma_S(k_p,m)\sigma_S(k_r,k_p)&=e(m,k_p)^{-1}\sigma_S(k_r,k_p)=e(m,k_p)^{-1}\min_{j\in I_r}\frac{e(j,k_p)}{e(j,k_r)}\\&\le e(m,k_p)^{-1}\frac{e(j_0,k_p)}{e(j_0,k_r)}\le\frac{e(j_0,m)}{e(j_0,k_r)}.\end{aligned}$$

Taking the minimum over $j_0\in I_r$ it follows that $\sigma_S(k_r,m)\ge\sigma_S(k_r,k_p)\sigma_S(k_p,m)$, which establishes the first inequality in (b) and completes the proof. □

LEMMA 3.9. *Let $S=\big((I_1,k_1);\dots;(I_s,k_s)\big)\in\mathcal{S}(n)$.*

(a) *Let $\mathbf{t}\in\mathbb{B}(1)$. If $\mathbf{t}\in E_S^A$ then for every $r\in\{1,\dots,s\}$*

$$\begin{aligned}&n_{k_r}(\mathbf{t}_{k_r})>A^{-1}n_j(\mathbf{t}_j)^{1/e(j,k_r)}=A^{-1}n_j(\mathbf{t}_j)^{\sigma_S(k_r,j)} &&\forall j\in I_r, j\ne k_r,\quad\text{and}\\&n_{k_r}(\mathbf{t}_{k_r})>A^{-1}n_{k_p}(\mathbf{t}_{k_p})^{\sigma_S(k_r,k_p)} &&\forall p\in\{1,\dots,s\},\ p\ne r.\end{aligned}\tag{3.1}$$

Conversely, if $\mathbf{t}\in\mathbb{B}(1)$ satisfies (3.1) then $\mathbf{t}\in E_S^{A^\eta}$ where η is a constant that depends only on the matrix $\mathbf{E}$.

(b) *Let $\mathbf{t}\in\mathbb{B}(1)^c$. If $\mathbf{t}\in\widehat{E}_S^A$ then for every $r\in\{1,\dots,s\}$*

$$\begin{aligned}&n_{k_r}(\mathbf{t}_{k_r})>A^{-1}n_j(\mathbf{t}_j)^{e(k_r,j)}=A^{-1}n_j(\mathbf{t}_j)^{1/\tau_S(j,k_r)} &&\forall j\in I_r, j\ne k_r,\quad\text{and}\\&n_{k_r}(\mathbf{t}_{k_r})>A^{-1}n_{k_p}(\mathbf{t}_{k_p})^{1/\tau_S(k_p,k_r)} &&\forall p\in\{1,\dots,s\},\ p\ne r.\end{aligned}\tag{3.2}$$

Conversely, if $\mathbf{t}\in\mathbb{B}(1)^c$ satisfies (3.2) then $\mathbf{t}\in E_S^{A^\eta}$ where η is a constant that depends only on the matrix $\mathbf{E}$.

Letting $A=A^\eta=1$, we obtain the following characterization of the sets E_S and $\widehat{E}_S$.

COROLLARY 3.10. Let $S=\big((I_1,k_1);\dots;(I_s,k_s)\big)\in\mathcal{S}(n)$.

(a) If $\mathbf{t}\in\mathbb{B}(1)$ then $\mathbf{t}\in E_S$ if and only if

$$\begin{aligned}&n_{k_r}(\mathbf{t}_{k_r})>n_j(\mathbf{t}_j)^{1/e(j,k_r)}=n_j(\mathbf{t}_j)^{\sigma_S(k_r,j)} &&\text{for every } j\in I_r \text{ with } j\ne k_r,\text{ and}\\&n_{k_r}(\mathbf{t}_{k_r})>n_{k_p}(\mathbf{t}_{k_p})^{\sigma_S(k_r,k_p)} &&\text{for every } p\in\{1,\dots,s\}\text{ with } p\ne r.\end{aligned}$$

(b) If $\mathbf{t} \in \mathbb{B}(1)^c$ then $\mathbf{t} \in \widehat{E}_S$ if and only if

$$n_{k_r}(\mathbf{t}_{k_r}) > n_j(\mathbf{t}_j)^{e(k_r,j)} = n_j(t_j)^{1/\tau_S(j,k_r)} \quad \text{for every } j \in I_r \text{ with } j \neq k_r, \text{ and}$$
$$n_{k_r}(\mathbf{t}_{k_r}) > n_{k_p}(\mathbf{t}_{k_p})^{1/\tau_S(k_p,k_r)} \quad \text{for every } p \in \{1,\ldots,s\} \text{ with } p \neq r.$$

PROOF OF LEMMA 3.9.

If $\mathbf{t} \in \mathbb{B}(1)$ then $\mathbf{t} \in E_S^A$ if and only if $n_{k_r}(\mathbf{t}_{k_r}) > A^{-1}n_l(\mathbf{t}_l)^{e(j,l)/e(j,k_r)}$ for $1 \leq r \leq s$, for $l \neq k_r$, and for all $j \in I_r$. But since $n_l(\mathbf{t}_l) \leq 1$ this is equivalent to saying that

$$\begin{aligned} n_{k_r}(\mathbf{t}_{k_r}) &> A^{-1}\max_{j\in I_r} n_l(\mathbf{t}_l)^{e(j,l)/e(j,k_r)} \\ &= A^{-1}n_l(\mathbf{t}_l)^{\min_{j\in I_r} e(j,l)/e(j,k_r)} \\ &= A^{-1}n_l(\mathbf{t}_l)^{\sigma_S(k_r,l)}, \end{aligned}$$

and this establishes both inequalities in (3.1). To prove the converse, let $l \in \{1,\ldots,n\}$ with $l \neq k_r$. Then there exists an index $p \in \{1,\ldots,s\}$ so that $l \in I_p$, and by the first inequality in (3.1) it follows that $n_l(\mathbf{t}_l)^{\sigma_S(k_p,l)} < A\,n_{k_p}(\mathbf{t}_{k_p})$. On the other hand, the second inequality in (3.1) says that $n_{k_p}(\mathbf{t}_{k_p})^{\sigma_S(k_r,k_p)} \leq A\,n_{k_r}(\mathbf{t}_{k_r})$. Also, Proposition 3.8, part (b) says that $\sigma_S(k_r,l) \geq \sigma_S(k_r,k_p)\sigma_S(k_p,l)$. Since $n_l(\mathbf{t}_l) < 1$, it follows that

$$\begin{aligned} n_l(\mathbf{t}_l)^{\sigma_S(k_r,l)} &\leq n_l(\mathbf{t}_l)^{\sigma_S(k_p,l)\sigma_S(k_r,k_p)} \\ &< \left[A\,n_{k_p}(\mathbf{t}_{k_t})\right]^{\sigma_S(k_r,k_p)} < A^{1+\sigma_S(k_r,k_p)}n_{k_r}(\mathbf{t}_{k_r}). \end{aligned}$$

But then if $j \in I_r$ and $l \neq k_r$,

$$\begin{aligned} n_l(\mathbf{t}_l)^{e(j,l)} &= \left[n_l(\mathbf{t}_l)^{e(j,l)/e(j,k_r)}\right]^{e(j,k_r)} \leq n_l(\mathbf{t}_l)^{\sigma_S(k_r,l)e(j,k_r)} \\ &< \left[A^{(1+\sigma_S(k_r,k_p))}n_{k_r}(\mathbf{t}_{k_r})\right]^{e(j,k_r)} \leq \left[A^{\eta}n_{k_r}(\mathbf{t}_{k_r})\right]^{e(j,k_r)} \end{aligned}$$

where we take $\eta = 1 + \sup_{r,p}\sigma_S(k_r,k_p) + \sup_{r,p}\tau_S(k_p,k_r)^{-1}$. This shows that $\mathbf{t}_{k_r}$ is A^{η}-dominant in $N_j(\mathbf{t})$, and completes the proof of (a).

The proof of (b) is very similar. Suppose that $\mathbf{t} \in \mathbb{B}(1)^c$. Then $\mathbf{t} \in \widehat{E}_S^A$ if and only if $n_{k_r}(\mathbf{t}_{k_r}) > A^{-1}\,n_l(\mathbf{t}_l)^{e(k_r,j)/e(l,j)}$ for $1 \leq r \leq s$, for $l \neq k_r$, and for all $j \in I_r$. Suppose first that $n_l(\mathbf{t}_l) \geq 1$. In this case it follows that

$$\begin{aligned} n_{k_r}(\mathbf{t}_{k_r}) &> A^{-1}\max_{j\in I_r} n_l(\mathbf{t}_l)^{e(k_r,j)/e(l,j)} \\ &= A^{-1}n_l(\mathbf{t}_l)^{\max_{j\in I_r} e(k_r,j)/e(l,j)} = A^{-1}n_l(\mathbf{t}_l)^{1/\tau_S(l,k_r)}. \end{aligned}$$

On the other hand, if $n_l(\mathbf{t}_l) < 1$ then we still have $n_{k_r}(\mathbf{t}_{k_r}) > A^{-1}n_l(\mathbf{t}_l)^{1/\tau_S(l,k_r)}$ since $n_{k_r}(\mathbf{t}_{k_r}) > A^{-1}$ by Remarks 3.2. In either case, this establishes both inequalities of (3.2). To prove the converse, let $l \in \{1,\ldots,n\}$ with $l \neq k_r$. If $n_l(\mathbf{t}_l) < 1$, then since $n_{k_r}(\mathbf{t}_{k_r}) > A^{-1}$ it follows that $n_l(\mathbf{t}_l)^{1/e(l,j)} < 1 < \left[A\,n_{k_r}(\mathbf{t}_{k_r})\right]^{1/e(k_r,j)}$. If $n_l(\mathbf{t}_l) \geq 1$. There exists $p \in \{1,\ldots,s\}$ so that $l \in I_p$, and by the first inequality in (3.2) it follows that $n_l(\mathbf{t}_l)^{1/\tau_S(l,k_p)} < A\,n_{k_p}(\mathbf{t}_{k_p})$. On the other hand, the second inequality in (3.2) says that $n_{k_p}(\mathbf{t}_{k_p}) \leq A^{\tau_S(k_p,k_r)}\,n_{k_r}(\mathbf{t}_{k_r})$. Also, Proposition 3.8, part (b) says that $\tau_S(l,k_r) \geq \tau_S(l,k_p)\tau_S(k_p,k_r)$. It follows that

$$\begin{aligned} n_l(\mathbf{t}_l)^{1/\tau_S(l,k_r)} &\leq n_l(\mathbf{t}_l)^{1/\tau_S(l,k_p)\tau_S(k_p,k_r)} \\ &< \left[A\,n_{k_p}(\mathbf{t}_{k_p})\right]^{1/\tau_S(k_p,k_r)} < A^{1+1/\tau_S(k_p,k_r)}n_{k_r}(\mathbf{t}_{k_r}). \end{aligned}$$

If $j \in I_r$ and $l \neq k_r$,

$$n_l(\mathbf{t}_l)^{1/e(l,j)} \leq \big[n_l(\mathbf{t}_l)^{1/\tau_S(l,k_r)}\big]^{1/e(k_r,j)} < \big[A^{1+1/\tau_S(k_p,k_r)} n_{k_r}(\mathbf{t}_{k_r})\big]^{1/e(k_r,j)}.$$

Thus in all cases we have $n_l(\mathbf{t}_l)^{1/e(k_r,l)} < \big[A^\eta\, n_{k_r}(\mathbf{t}_{k_r})\big]^{1/e(k_r,j)}$, and this says that $\mathbf{t}_{k_r}$ is A^η-dominant in $\widehat{N}_j(\mathbf{t})$, which completes the proof of (b) and of Lemma 3.9. □

3.4. Estimates of kernels and multipliers on E_S and $\widehat{E}_S$

The differential inequalities for kernels and multipliers take simpler forms on the sets E_S and $\widehat{E}_S$. First consider the principal marked partition

$$S_0 = \big\{(\{1\},1);\ldots;(\{n\},n)\big\}.$$

If $\mathbf{x} \in \mathbb{B}(1) \cap E_{S_0}$ then $N_k(\mathbf{x}) \approx n_k(\mathbf{x}_k)$, and if $\boldsymbol{\xi} \in \widehat{E}_{S_0}$ then $\widehat{N}_k(\boldsymbol{\xi}) \approx n_k(\boldsymbol{\xi}_k)$. Thus the differential inequalities for $\mathcal{K} \in \mathcal{P}(\mathbf{E})$ and $m \in \mathcal{M}(\mathbf{E})$ take the following form on the sets E_{S_0} and $\widehat{E}_{S_0}$:

$$\begin{aligned} \mathbf{x} \in E_{S_0} \quad &\Longrightarrow \quad \big|\partial^\gamma K(\mathbf{x})\big| \leq C_\gamma \prod_{k=1}^n n_k(\mathbf{x}_k)^{-(Q_k + [\![\gamma_k]\!])}, \\ \boldsymbol{\xi} \in \widehat{E}_{S_0} \quad &\Longrightarrow \quad \big|\partial^\gamma m(\boldsymbol{\xi})\big| \leq C_\gamma \prod_{k=1}^n n_k(\boldsymbol{\xi}_k)^{-[\![\gamma_k]\!]}. \end{aligned}$$

Note that these are precisely the differential inequalities satisfied by product kernels or product multipliers on $\mathbb{R}^N = \bigoplus_{j=1}^n \mathbb{R}^{C_j}$. (See [**NRS01**], page 34 and page 37.) Thus although a distribution $\mathcal{K} \in \mathcal{P}(\mathbf{E})$ is pseudo-local, if the set $E_{S_0} \neq \emptyset$ the function K satisfies nothing better than the estimates for a product kernel on this set.

We have the following extension for non-empty sets E_S and $\widehat{E}_S$.

PROPOSITION 3.11. Let $S = \big\{(I_1, k_1);\ldots;(I_s;k_s)\big\} \in \mathcal{S}(n)$.

(a) If $E_S \cap \mathbb{B}(1) \neq \emptyset$, if $\mathcal{K} \in \mathcal{P}(\mathbf{E})$, and if $K \in \mathcal{C}^\infty(\mathbb{R}^N \setminus \{0\})$ is the associated smooth function, then for every $\boldsymbol{\gamma} = (\gamma_1,\ldots,\gamma_n)$, there is a constant C_γ so that for $\mathbf{x} \in E_S \cap \mathbb{B}(1)$ the inequality in Definition 2.2, part (a) is equivalent to

$$\big|\partial^\gamma K(\mathbf{x})\big| \leq C_\gamma \prod_{r=1}^s n_{k_r}(\mathbf{x}_{k_r})^{-\sum_{j\in I_r} e(j,k_r)(Q_j + [\![\gamma_j]\!])}.$$

(b) If $\widehat{E}_S \cap \mathbb{B}(1)^c \neq \emptyset$ and if $m \in \mathcal{M}(\mathbf{E})$ then for every $\boldsymbol{\gamma} = (\gamma_1,\ldots,\gamma_n)$ there is a constant C_γ so that for $\boldsymbol{\xi} \in \widehat{E}_S$ the inequality in Definition 2.4 is equivalent to

$$\big|\partial^\gamma m(\boldsymbol{\xi})\big| \leq C_\gamma \prod_{r=1}^s n_{k_r}(\boldsymbol{\xi}_{k_r})^{-\sum_{j\in I_r} e(j,k_r)[\![\gamma_j]\!]}.$$

PROOF. If $\mathbf{x} \in E_S$, then $N_j(\mathbf{x}) \approx n_{k_r}(\mathbf{x}_{k_r})^{e(j,k_r)}$ for every $j \in I_r$. Thus if $\mathbf{x} \in E_S$

$$\big|\partial^\gamma K(\mathbf{x})\big| \leq C_\gamma \prod_{j=1}^n N_j(\mathbf{x})^{-(Q_j + [\![\gamma_j]\!])} \approx C_\gamma \prod_{r=1}^s \prod_{j\in I_r} n_{k_r}(\mathbf{x}_{k_r})^{-e(j,k_r)(Q_j + [\![\gamma_j]\!])}.$$

This establishes part (a), and again part (b) is done the same way. □

3.5. A coarser decomposition of $\mathbb{R}^N$ associated to $S \in \mathcal{S}(n)$

Proposition 3.11 suggests that, on each non-empty set E_S, a kernel $\mathcal{K} \in \mathcal{P}_0(\mathbf{E})$ satisfies the same differential inequalities of a product kernel adapted to a coarser decomposition of $\mathbb{R}^N$ depending on S and, in a similar way, that on each non-empty set $\widehat{E}_S$, a multiplier $m \in \mathcal{M}_\infty(\mathbf{E})$ behaves like a product multiplier.

For each $1 \leq r \leq s$ we gather the coordinates $\{\mathbf{t}_j : j \in I_r\}$ into a single subspace as discussed on page 11. Thus for $1 \leq r \leq s$ set

$$\mathbb{R}^{I_r} = \bigoplus_{j \in I_r} \mathbb{R}^{C_j} \tag{3.3}$$

so that $\mathbb{R}^N = \mathbb{R}^{I_1} \oplus \cdots \oplus \mathbb{R}^{I_s}$. Note that if $s < n$, this decomposition is strictly coarser than the original decomposition $\mathbb{R}^N = \mathbb{R}^{C_1} \oplus \cdots \oplus \mathbb{R}^{C_n}$. If $\mathbf{t} \in \mathbb{R}^N$ write $\mathbf{t} = (\mathbf{t}_{I_1}, \ldots, \mathbf{t}_{I_S})$ where $\mathbf{t}_{I_r} = \big\{\mathbf{t}_j : j \in I_r\big\}$.

Next we introduce two new families of dilations $\mathbf{x}_{I_r} \to \lambda \cdot_S \mathbf{x}_{I_r}$ and $\boldsymbol{\xi}_{I_r} \to \lambda \hat{\cdot}_S \boldsymbol{\xi}_{I_r}$ on each component $\mathbb{R}^{I_r}$. These should have the property that if $\mathbf{x} \in E_S$ then the associated norm of $\mathbf{x}_{I_r}$ should be comparable to $n_{k_r}(\mathbf{x}_{k_r})$, and if $\boldsymbol{\xi} \in \widehat{E}_S$, then the associated norm of $\boldsymbol{\xi}_{I_r}$ should be comparable to $n_{k_r}(\boldsymbol{\xi}_{k_r})$. Recall from Corollary 3.10 that

$$\begin{aligned} \mathbf{x} \in E_S \quad &\Longrightarrow \quad n_j(\mathbf{x}_j)^{1/e(j,k_r)} < n_{k_r}(\mathbf{x}_{k_r}) \quad \text{if } j \in I_r \text{ and } j \neq k_r, \\ \boldsymbol{\xi} \in \widehat{E}_S \quad &\Longrightarrow \quad n_j(\boldsymbol{\xi}_j)^{\,e(k_r,j)} < n_{k_r}(\boldsymbol{\xi}_{k_r}) \quad \text{if } j \in I_r \text{ and } j \neq k_r, \end{aligned} \tag{3.4}$$

and from Proposition 3.8 that if $j \in I_r$ then $e(j,k_r) = \sigma_S(k_r,j)^{-1}$ and $e(k_r,j) = \tau(j,k_r)^{-1}$.

DEFINITION 3.12. Let $S = \big\{(I_1,k_1); \ldots; (I_s,k_s)\big\} \in \mathcal{S}(n)$.

(a) If $\mathbf{x} = (\mathbf{x}_{I_1}, \ldots, \mathbf{x}_{I_s}) \in \mathbb{R}^N$ with $\mathbf{x}_{I_r} = \big\{\mathbf{x}_j : j \in I_r\big\} \in \mathbb{R}^{I_r}$ set

$$\lambda \cdot_S \mathbf{x}_{I_r} = \Big(\lambda^{e(j,k_r)} \cdot \mathbf{x}_j : j \in I_r\Big) = \Big(\lambda^{1/\sigma_S(k_r,j)} \cdot \mathbf{x}_j : j \in I_r\Big).$$

Let $n_{S,r}$ be a smooth homogeneous norm on $\mathbb{R}^{I_r}$ for this dilation.

(b) If $\boldsymbol{\xi} = (\boldsymbol{\xi}_{I_1}, \ldots, \boldsymbol{\xi}_{I_s}) \subset \mathbb{R}^N$ with $\boldsymbol{\xi}_{I_r} - \big\{\boldsymbol{\xi}_j : j \in I_r\big\}$ let

$$\lambda \hat{\cdot}_S \boldsymbol{\xi}_{I_r} = \Big(\lambda^{1/e(k_r,j)} \cdot \boldsymbol{\xi}_j : j \in I_r\Big) = \Big(\lambda^{\tau_S(j,k_r)} \cdot \boldsymbol{\xi}_j : j \in I_r\Big).$$

Let $\widehat{n}_{S,r}$ be a smooth homogeneous norm on $\mathbb{R}^{I_r}$ for this dilation.

Observe that

$$\begin{aligned} n_{S,r}(\mathbf{x}_{I_r}) &\approx \sum_{j \in I_r} n_j(\mathbf{x}_j)^{1/e(j,k_r)} = \sum_{j \in I_r} n_j(\mathbf{x}_j)^{\sigma_S(k_r,j)}, \\ \widehat{n}_{S,r}(\boldsymbol{\xi}_{I_r}) &\approx \sum_{j \in I_r} n_j(\boldsymbol{\xi}_j)^{e(k_r,j)} = \sum_{j \in I_r} n_j(\boldsymbol{\xi}_j)^{1/\tau_S(j,k_r)}, \end{aligned} \tag{3.5}$$

and that the homogeneous dimensions of $\mathbb{R}^{I_r}$ under the dilations $\lambda \cdot_S$ and $\lambda \hat{\cdot}_S$ are given by

$$
\begin{aligned}
Q_{S,r} &= \sum_{j \in I_r} Q_j e(j, k_r) = \sum_{j \in I_r} Q_j \sigma_S(k_r, j)^{-1}, \\
\widehat{Q}_{S,r} &= \sum_{j \in I_r} Q_j e(k_r, j)^{-1} = \sum_{j \in I_r} Q_j \tau_S(j, k_r).
\end{aligned} \tag{3.6}
$$

Moreover, it follows directly from and (2.9) and (3.5) that, for $1 \le r \le s$,

$$
\begin{aligned}
n_{S,r}(\mathbf{t}_{I_r}) &\lesssim \widehat{N}_{k_r}(\mathbf{t}), \\
\widehat{n}_{S,r}(\mathbf{t}_{I_r}) &\lesssim N_{k_r}(\mathbf{t}),
\end{aligned} \tag{3.7}
$$

for all $\mathbf{t} \in \mathbb{R}^N$.

CHAPTER 4

Fourier transform duality of kernels and multipliers

In this chapter we show that the Fourier transforms of kernels $\mathcal{K} \in \mathcal{P}_0(\mathbb{E})$ are multipliers $m \in \mathcal{M}_\infty(\mathbf{E})$, and conversely, that the inverse Fourier transform of multipliers $m \in \mathcal{M}_\infty(\mathbf{E})$ are kernels $\mathcal{K} \in \mathcal{P}_0(\mathbf{E})$. Such results are well known for Calderón-Zygmund kernels and multipliers, and more generally, for product kernels and product multipliers or for flag kernels and flag multipliers. (See [**NRS01**] for details.)

4.1. Fourier transforms of multipliers

Let $m \in L^\infty(\mathbb{R}^N)$ and let $\mathcal{K} = \widehat{m}$ be its Fourier transform in the sense of tempered distributions. Thus if $\varphi \in \mathcal{S}(\mathbb{R}^N)$ we have $\langle \mathcal{K}, \varphi \rangle = \langle m, \widehat{\varphi} \rangle = \int_{\mathbb{R}^N} m(\boldsymbol{\xi})\widehat{\varphi}(\boldsymbol{\xi})\,d\boldsymbol{\xi}$. Let $\chi \in \mathcal{C}_0^\infty(\mathbb{R}^N)$ be identically equal to one in a neighborhood of the origin, and let $\chi_\epsilon(\boldsymbol{\xi}) = \chi(\epsilon\boldsymbol{\xi})$. Then

$$\langle \mathcal{K}, \varphi \rangle = \int_{\mathbb{R}^N} m(\boldsymbol{\xi})\widehat{\varphi}(\boldsymbol{\xi})\,d\boldsymbol{\xi} = \lim_{\epsilon\to 0} \int_{\mathbb{R}^N} \chi_\epsilon(\boldsymbol{\xi})m(\boldsymbol{\xi})\widehat{\varphi}(\boldsymbol{\xi})\,d\boldsymbol{\xi} = \lim_{\epsilon\to 0} \langle \widehat{\chi_\epsilon m}, \varphi \rangle$$

so $\mathcal{K} = \lim_{\epsilon\to 0} \widehat{\chi_\epsilon m}$ in the sense of distributions. Thus in making estimates of the Fourier transform $\widehat{m}$, we can assume that m has compact support (and hence can freely integrate by parts) provided that we do not use any information about the size of this support in our estimates.

THEOREM 4.1. *Let $m \in \mathcal{M}_\infty(\mathbf{E})$. Then the Fourier transform $\widehat{m}$ is a distribution $\mathcal{K}$ belonging to the class $\mathcal{P}_0(\mathbf{E})$.*

PROOF. Assume as discussed above that m has compact support. The Fourier transform is given by $K(\mathbf{x}) = \widehat{m}(\mathbf{x}) = \int_{\mathbb{R}^N} e^{2\pi i\langle \mathbf{x},\boldsymbol{\xi}\rangle} m(\boldsymbol{\xi})\,d\boldsymbol{\xi}$, and for every $\boldsymbol{\gamma} \in \mathbb{N}^N$

$$\partial^\gamma K(\mathbf{x}) = (2\pi i)^{|\gamma|} \int_{\mathbb{R}^N} e^{2\pi i\langle \mathbf{x},\boldsymbol{\xi}\rangle}\, \boldsymbol{\xi}^\gamma\, m(\boldsymbol{\xi})\,d\boldsymbol{\xi}. \tag{4.1}$$

To show that $\mathcal{K} = \widehat{m} \in \mathcal{P}_0(\mathbf{E})$, we must verify the differential inequalities of Definition 2.2 and Definition 2.5 as well as the the cancellation conditions of Definition 2.2. We begin with the differential inequalities.

If $\boldsymbol{\gamma} \in \mathbb{N}^{C_1} \times \cdots \times \mathbb{N}^{C_n}$ and $M \geq 0$, we must show that there are constants C_γ, $C_{\gamma,M} > 0$ (independent of the support of m) so that

$$\begin{aligned} \big|\partial^\gamma K(\mathbf{x})\big| &\leq C_\gamma \prod_{j=1}^n N_j(\mathbf{x})^{-(Q_j+[\![\gamma_j]\!])} &&\text{if } \mathbf{x} \in \mathbb{B}(1), \\ \big|\partial^\gamma K(\mathbf{x})\big| &\leq C_{\gamma,M}(1+|\mathbf{x}|)^{-M} &&\text{if } \mathbf{x} \notin \mathbb{B}(1), \end{aligned} \tag{4.2}$$

where $|\mathbf{x}|$ is the ordinary Euclidean norm of $\mathbf{x} \in \mathbb{R}^N$.

The second inequality in (4.2) follows from a standard integration by parts. Let Δ denote the ordinary Laplace operator. Then

$$\begin{aligned}\int_{\mathbb{R}^N} e^{2\pi i\langle \mathbf{x},\boldsymbol{\xi}\rangle}\,\boldsymbol{\xi}^\gamma\, m(\boldsymbol{\xi})\,d\boldsymbol{\xi} &= (4\pi^2|\mathbf{x}|)^{-2M}\int_{\mathbb{R}^N}(-\Delta_{\boldsymbol{\xi}})^M\left[e^{2\pi i\langle \mathbf{x},\boldsymbol{\xi}\rangle}\right]\boldsymbol{\xi}^\gamma\, m(\boldsymbol{\xi})\,d\boldsymbol{\xi}\\ &= (4\pi^2|\mathbf{x}|)^{-2M}\int_{\mathbb{R}^N} e^{2\pi i\langle \mathbf{x},\boldsymbol{\xi}\rangle}(-\Delta_{\boldsymbol{\xi}})^M\left[\boldsymbol{\xi}^\gamma\, m(\boldsymbol{\xi})\right]d\boldsymbol{\xi}.\end{aligned}$$

If M is large enough, $\left|\Delta_{\boldsymbol{\xi}}^M\left[\boldsymbol{\xi}^\gamma\, m(\boldsymbol{\xi})\right]\right| \le C_M\big(1+|\boldsymbol{\xi}|\big)^{-N-1}$, so the last integral converges absolutely independently of the support of m.

We turn to the heart of the matter which is the first estimate in (4.2). Suppose that $\mathbf{x}\in\mathbb{B}(1)\cap E_S$ where $S=\big\{(I_1,k_1);\dots;(I_s,k_s)\big\}\in\mathcal{S}(n)$. By (3.4), for $1\le r\le s$, if $j\in I_r$ we have

$$N_j(\mathbf{x})\approx n_{k_r}(\mathbf{x}_{k_r})^{e(j,k_r)}\approx n_{S,r}(\mathbf{x}_{I_r})^{e(j,k_r)},\tag{4.3}$$

where $n_{S,r}(\mathbf{x}_{I_r})\approx\sum_{j\in I_r}n_j(\mathbf{x}_j)^{1/e(j,k_r)}$ is the norm defined in (3.5). Hence, for $\mathbf{x}\in\mathbb{B}(1)\cap E_S$, the estimate (4.2) is equivalent to the estimate

$$\big|\partial^\gamma K(\mathbf{x})\big|\lesssim\prod_{r=1}^s n_{k_r}(\mathbf{x}_{k_r})^{-\sum_{j\in I_r}e(j,k_r)(Q_j+[\![\gamma_j]\!])},\tag{4.4}$$

which, by (4.3), may also be interpreted as the differential inequality of a product kernel for the decomposition $\mathbb{R}^N=\mathbb{R}^{I_1}\oplus\cdots\oplus\mathbb{R}^{I_s}$ with the dilations $\cdot_S$ such that $\lambda\cdot_S\mathbf{x}_{I_r}=\big\{\lambda^{e(j,k_r)}\cdot\mathbf{x}_j: j\in I_r\big\}$.

To obtain (4.4) we adapt the standard method used for proving that product kernels and multipliers are related via the Fourier transform, cf. Theorem 2.1.11 in [**NRS01**]. This consists in splitting the Fourier integral (4.1) into 2^s regions depending on whether $n_{S,r}(\boldsymbol{\xi}_{I_r})$ is larger or smaller[1] than $n_{k_r}(\mathbf{x}_{k_r})^{-1}$ for $1\le r\le s$. For any r, we take advantage of the smallness of the domain of integration if we are considering $n_{S,r}(\boldsymbol{\xi}_{I_r})<n_{k_r}(\mathbf{x}_{k_r})^{-1}$, while we integrate by parts in $\boldsymbol{\xi}_{k_r}$ in the other case[2].

Let φ_0 and φ_1 be smooth functions on the positive half-line with $\varphi_0(t)+\varphi_1(t)\equiv 1$, $\varphi_0(t)\equiv 1$ for $t\le\frac12$ and $\varphi_0(t)\equiv 0$ for $t\ge 2$. For $\epsilon=(\epsilon_1,\dots,\epsilon_s)\in\{0,1\}^s$ set

$$\begin{aligned}m_\epsilon(\boldsymbol{\xi})&=m(\boldsymbol{\xi})\prod_{r=1}^s\varphi_{\epsilon_r}\big(n_{S,r}(\boldsymbol{\xi}_{I_r})\,n_{k_r}(\mathbf{x}_{k_r})\big),\\ K_\epsilon(\mathbf{x})&=\int_{\mathbb{R}^N}e^{2\pi i\langle\mathbf{x},\boldsymbol{\xi}\rangle}m_\epsilon(\boldsymbol{\xi})\,d\boldsymbol{\xi}.\end{aligned}$$

[1] It may seem strange that we measure the size of $\boldsymbol{\xi}_{I_r}$ with $n_{S,r}(\boldsymbol{\xi}_{I_r})$ instead of $\widehat{n}_{s,r}(\boldsymbol{\xi}_{I_r})$. The point is that we have estimates of derivatives of m in terms of $\widehat{N}_{k_r}(\boldsymbol{\xi})$ and we want a bound from below on this quantity that involves a norm on $\mathbb{R}^{I_r}$. By (3.7), we need to use $n_{S,r}(\boldsymbol{\xi}_{I_r})$.

[2] We are not claiming that m is a product multiplier for the coarser decomposition. If it were so, the estimate

$$\big|\partial^\gamma K(\mathbf{x})\big|\lesssim\prod_{r=1}^s n_{S,r}(\mathbf{x}_{I_r})^{-\sum_{j\in I_r}e(j,k_r)(Q_j+[\![\gamma_j]\!])},$$

would hold for every $\mathbf{x}\in\mathbb{R}^N$, which is not true. Also notice that formula (4.6) below, which gives a differential inequality typical of a product multiplier, only holds for derivatives in the $\boldsymbol{\xi}_{k_r}$.

Note that $m(\boldsymbol{\xi}) = \sum_{\epsilon\in\{0,1\}^s} m_{\boldsymbol{\epsilon}}(\boldsymbol{\xi})$ and $K(\mathbf{x}) = \sum_{\epsilon\in\{0,1\}^s} K_\epsilon(\mathbf{x})$, so it will suffice to show that

$$\big|\partial^\gamma K_\epsilon(\mathbf{x})\big| \lesssim \prod_{r=1}^s n_{k_r}(\mathbf{x}_{k_r})^{-\sum_{j\in I_r} e(j,k_r)(Q_j+[\![\boldsymbol{\gamma}_j]\!])}. \tag{4.5}$$

Since $m \in \mathcal{M}(\mathbf{E})$, and since $\widehat{N}_{k_r}(\boldsymbol{\xi}) \gtrsim n_{S,r}(\boldsymbol{\xi}_{I_r})$ by (3.7), we have

$$\big|\partial^{\boldsymbol{\delta}_{k_1}}_{\boldsymbol{\xi}_{k_1}} \cdots \partial^{\boldsymbol{\delta}_{k_s}}_{\boldsymbol{\xi}_{k_s}} m(\boldsymbol{\xi})\big| \lesssim \prod_{r=1}^s \widehat{N}_{k_r}(\boldsymbol{\xi})^{-[\![\boldsymbol{\delta}_{k_r}]\!]} \lesssim \prod_{r=1}^s n_{S,r}(\boldsymbol{\xi}_{I_r})^{-[\![\boldsymbol{\delta}_{k_r}]\!]}. \tag{4.6}$$

Since $n_{S,r}(\boldsymbol{\xi}_{I_r})$ is homogeneous of degree one under the dilations $\lambda \cdot_S \boldsymbol{\xi}_{I_r}$, and since $e(k_r, k_r) = 1$, it follows from Proposition 15.1 in Appendix B that for all multi-indices γ

$$\big|\partial^\gamma_{\boldsymbol{\xi}_{k_r}} n_{S,r}(\boldsymbol{\xi}_{I_r})\big| \leq C_\gamma n_{S,r}(\boldsymbol{\xi}_{I_r})^{1-[\![\gamma]\!]}$$

Using the chain rule, we see that $\partial^{\boldsymbol{\delta}_{k_r}}_{\boldsymbol{\xi}_{k_r}}\Big[\varphi_{\epsilon_r}\big(n_{S,r}(\boldsymbol{\xi}_{I_r})\, n_{k_r}(\mathbf{x}_{k_r})\big)\Big]$ is a sum of terms of the form

$$\varphi^{(m)}_{\epsilon_r}\big(n_{S,r}(\boldsymbol{\xi}_{I_r}) n_{k_r}(\mathbf{x}_{k_r})\big)\, n_{k_r}(\mathbf{x}_{k_r})^m\, \partial^{\boldsymbol{\alpha}_1}_{\boldsymbol{\xi}_{k_r}} n_{S,r}(\boldsymbol{\xi}_{I_r}) \cdots \partial^{\boldsymbol{\alpha}_m}_{\boldsymbol{\xi}_{k_r}} n_{S,r}(\boldsymbol{\xi}_{I_r})$$

where $1 \leq m \leq |\boldsymbol{\delta}_{k_r}|$ and $\boldsymbol{\alpha}_1 + \cdots + \boldsymbol{\alpha}_m = \delta_{k_r}$. Since $n_{k_r}(\mathbf{x}_{k_r}) \approx n_{S,r}(\boldsymbol{\xi}_{I_r})^{-1}$ on the support of $\varphi^{(m)}_{\epsilon_r}$, $m \geq 1$, it follows that

$$\Big|\varphi^{(m)}_{\epsilon_r}\big(n_{S,r}(\boldsymbol{\xi}_{I_r}) n_{k_r}(\mathbf{x}_{k_r})\big)\, n_{k_r}(\mathbf{x}_{k_r})^m\, \partial^{\boldsymbol{\alpha}_1}_{\boldsymbol{\xi}_{k_r}} n_{S,r}(\boldsymbol{\xi}_{I_r}) \cdots \partial^{\boldsymbol{\alpha}_m}_{\boldsymbol{\xi}_{k_r}} n_{S,r}(\boldsymbol{\xi}_{I_r})\Big| \lesssim n_{S,r}(\boldsymbol{\xi}_{I_r})^{-[\![\boldsymbol{\delta}_{k_r}]\!]}.$$

Combining these estimates gives $\big|\partial^{\boldsymbol{\delta}_{k_1}}_{\boldsymbol{\xi}_{k_1}} \cdots \partial^{\boldsymbol{\delta}_{k_s}}_{\boldsymbol{\xi}_{k_s}} m_\epsilon(\boldsymbol{\xi})\big| \leq C_{\boldsymbol{\delta}} \prod_{r=1}^s n_{S,r}(\boldsymbol{\xi}_{I_r})^{-[\![\boldsymbol{\delta}_{k_r}]\!]}$.

Now for each $\epsilon = (\epsilon_1, \ldots, \epsilon_s) \in \{0,1\}^s$ let $A(\epsilon) = \{r \in \{1, \ldots, s\} : \epsilon_r = 1\}$. Note that m_ϵ is supported where

$$\begin{aligned} n_{S,r}(\boldsymbol{\xi}_{I_r}) &\leq 2n_{k_r}(\mathbf{x}_{k_r})^{-1} && \text{if } r \notin A(\epsilon), \\ n_{S,r}(\boldsymbol{\xi}_{I_r}) &\geq \frac{1}{2} n_{k_r}(\mathbf{x}_{k_r})^{-1} && \text{if } r \in A(\epsilon). \end{aligned}$$

Choose $\boldsymbol{\delta} = (\boldsymbol{\delta}_1, \ldots, \boldsymbol{\delta}_n) \in \mathbb{N}^{C_1} \times \cdots \times \mathbb{N}^{C_n}$ with $\boldsymbol{\delta}_j = 0$ unless $j = k_r$ for some $r \in A(\epsilon)$. If $r \in A(\epsilon)$, we will choose the entries of each $\boldsymbol{\delta}_{k_r} = (\delta_{k_r,1}, \ldots, \delta_{k_r,c_{k_r}})$ to be much larger than the corresponding entries of $\boldsymbol{\gamma}_{k_r} = (\gamma_{k_r,1}, \ldots, \gamma_{k_r,c_{k_r}})$. Using (4.1) we integrate by parts to obtain

$$\begin{aligned} \Big(\prod_{r\in A(\boldsymbol{\epsilon})} \mathbf{x}_{k_r}^{\boldsymbol{\delta}_{k_r}}\Big)&\partial^\gamma_{\mathbf{x}} K_\epsilon(\mathbf{x}) \\ &= (2\pi i)^{|\gamma|} \int_{\mathbb{R}^N} \Big(\prod_{r\in A(\boldsymbol{\epsilon})} \mathbf{x}_{k_r}^{\boldsymbol{\delta}_{k_r}}\Big) e^{2\pi i\langle \mathbf{x},\boldsymbol{\xi}\rangle} \boldsymbol{\xi}^\gamma m_\epsilon(\boldsymbol{\xi})\, d\boldsymbol{\xi} \\ &= (2\pi i)^{|\gamma|-|\delta|} \int_{\mathbb{R}^N} \Big[\prod_{r\in A(\boldsymbol{\epsilon})} \partial^{\boldsymbol{\delta}_{k_r}}_{\boldsymbol{\xi}_{k_r}}\Big]\big(e^{2\pi i\langle \mathbf{x},\boldsymbol{\xi}\rangle}\big)\, \boldsymbol{\xi}^\gamma m_\epsilon(\boldsymbol{\xi})\, d\boldsymbol{\xi} \\ &= (-1)^{|\delta|}(2\pi i)^{|\gamma|-|\delta|} \int_{\mathbb{R}^N} e^{2\pi i\langle \mathbf{x},\boldsymbol{\xi}\rangle} \Big[\prod_{r\in A(\boldsymbol{\epsilon})} \partial^{\boldsymbol{\delta}_{k_r}}_{\boldsymbol{\xi}_{k_r}}\Big](\boldsymbol{\xi}^\gamma m_\epsilon)(\boldsymbol{\xi})\, d\boldsymbol{\xi}. \end{aligned} \tag{4.7}$$

The product rule shows that $\big[\prod_{r\in A(\epsilon)} \partial^{\boldsymbol{\delta}_{k_r}}_{\boldsymbol{\xi}_{k_r}}\big](\boldsymbol{\xi}^{\boldsymbol{\gamma}} m_{\epsilon})(\boldsymbol{\xi})$ can be written as a finite sum of terms of the form

$$\Big(\prod_{r\in A(\boldsymbol{\epsilon})} \partial^{\boldsymbol{\mu}_{k_r}}_{\boldsymbol{\xi}_{k_r}}\Big)[\boldsymbol{\xi}^{\boldsymbol{\gamma}}]\ \Big(\prod_{r\in A(\boldsymbol{\epsilon})} \partial^{\boldsymbol{\delta}_{k_r}-\boldsymbol{\mu}_{k_r}}_{\boldsymbol{\xi}_{k_r}}\Big)[m_{\epsilon}](\boldsymbol{\xi}) \tag{4.8}$$

where $\boldsymbol{\mu} = (\boldsymbol{\mu}_1, \ldots, \boldsymbol{\mu}_n) \le (\boldsymbol{\delta}_1, \ldots, \boldsymbol{\delta}_n)$. The first factor is zero if the entries of $\boldsymbol{\mu}_{k_r}$ are larger than the corresponding terms $\boldsymbol{\gamma}_{k_r}$. Thus our choice of $\boldsymbol{\delta}_{k_r}$ guarantees that the entries of $\boldsymbol{\delta}_{k_r} - \boldsymbol{\mu}_{k_r}$ are large enough to make the integrals below converge independently of the compact support of m. Note that $\boldsymbol{\mu}_j = \mathbf{0}$ unless $j = k_r$ for some $r \in A(\epsilon)$. Then

$$\Big(\prod_{r\in A(\boldsymbol{\epsilon})} \partial^{\boldsymbol{\mu}_{k_r}}_{\boldsymbol{\xi}_{k_r}}\Big)[\boldsymbol{\xi}^{\boldsymbol{\gamma}}] = C_{\boldsymbol{\mu},\boldsymbol{\gamma}} \prod_{r\in A(\epsilon)} \Big(\prod_{j\in I_r} \boldsymbol{\xi}_j^{\boldsymbol{\gamma}_j-\boldsymbol{\mu}_j}\Big) \prod_{r\notin A(\epsilon)} \Big(\prod_{j\in I_r} \boldsymbol{\xi}_j^{\boldsymbol{\gamma}_j}\Big)$$

According to (4.6) we also have

$$\Big|\Big(\prod_{r\in A(\boldsymbol{\epsilon})} \partial^{\boldsymbol{\delta}_{k_r}-\boldsymbol{\mu}_{k_r}}_{\boldsymbol{\xi}_{k_r}}\Big)[m_{\epsilon}](\boldsymbol{\xi})\Big| \le C_{\boldsymbol{\delta},\boldsymbol{\mu}} \prod_{r\in A(\epsilon)} n_{S,r}(\boldsymbol{\xi}_{I_r})^{-[\![\boldsymbol{\delta}_{k_r}]\!]+[\![\boldsymbol{\mu}_{k_r}]\!]}$$

Thus

$$\begin{aligned}\Big|\Big(\prod_{r\in A(\boldsymbol{\epsilon})} \partial^{\boldsymbol{\mu}_{k_r}}_{\boldsymbol{\xi}_{k_r}}\Big)[\boldsymbol{\xi}^{\boldsymbol{\gamma}}]\ &\Big(\prod_{r\in A(\boldsymbol{\epsilon})} \partial^{\boldsymbol{\delta}_{k_r}-\boldsymbol{\mu}_{k_r}}_{\boldsymbol{\xi}_{k_r}}\Big)[m_{\epsilon}](\boldsymbol{\xi})\Big| \\ &\le C_{\boldsymbol{\delta},\boldsymbol{\gamma}} \prod_{r\in A(\epsilon)} n_{S,r}(\boldsymbol{\xi}_{I_r})^{-[\![\boldsymbol{\delta}_{k_r}]\!]+[\![\boldsymbol{\mu}_{k_r}]\!]} \prod_{r\in A(\epsilon)} \Big(\prod_{j\in I_r} |\boldsymbol{\xi}_j^{\boldsymbol{\gamma}_j-\boldsymbol{\mu}_j}|\Big) \prod_{r\notin A(\epsilon)} \Big(\prod_{j\in I_r} |\boldsymbol{\xi}_j^{\boldsymbol{\gamma}_j}|\Big).\end{aligned} \tag{4.9}$$

Now using (4.7), (4.8), and (4.9), we can estimate $\big|\big(\prod_{r\in A(\boldsymbol{\epsilon})} \mathbf{x}_{k_r}^{\boldsymbol{\delta}_{k_r}}\big)\partial^{\boldsymbol{\gamma}}_{\mathbf{x}} K_{\boldsymbol{\epsilon}}(\mathbf{x})\big|$ by a finite sum of terms of the form

$$\begin{aligned}\prod_{r\in A(\epsilon)} \Bigg[\int_{n_{S,r}(\boldsymbol{\xi}_{I_r})\ge n_{k_r}(\mathbf{x}_{k_r})^{-1}} \Big(\prod_{j\in I_r} |\boldsymbol{\xi}_j^{\boldsymbol{\gamma}_j-\boldsymbol{\mu}_j}|\Big)\, n_{S,r}(\boldsymbol{\xi}_{I_r})^{-[\![\boldsymbol{\delta}_{k_r}]\!]+[\![\boldsymbol{\mu}_{k_r}]\!]}\, d\boldsymbol{\xi}_{I_r}\Bigg] \\ \times \prod_{r\notin A(\epsilon)} \Bigg[\int_{n_{S,r}(\boldsymbol{\xi}_{I_r})< n_{k_r}(\mathbf{x}_{k_r})^{-1}} \Big(\prod_{j\in I_r} |\boldsymbol{\xi}_j^{\boldsymbol{\gamma}_j}|\Big)\, d\boldsymbol{\xi}_{I_r}\Bigg].\end{aligned}$$

According to Proposition 15.3 in the appendix, this last expression is bounded by

$$\begin{aligned}\prod_{r\notin A(\epsilon)} n_{k_r}(\mathbf{x}_{k_r})^{-\sum_{j\in I_r} e(k_r,j)(Q_j+[\![\boldsymbol{\gamma}_j]\!])} \prod_{r\in A(\epsilon)} n_{k_r}(\mathbf{x}_{k_r})^{-\sum_{j\in I_r} e(k_r,j)(Q_j+[\![\boldsymbol{\gamma}_j]\!]-[\![\boldsymbol{\delta}_r]\!])} \\ = \prod_{r\in A(\epsilon)} n_{k_r}(\mathbf{x}_{k_r})^{[\![\boldsymbol{\delta}_r]\!]} \prod_{r=1}^{s} n_{k_r}(\mathbf{x}_{k_r})^{-\sum_{j\in I_r} e(k_r,j)(Q_j+[\![\boldsymbol{\gamma}_j]\!])}.\end{aligned}$$

Comparing with (4.7), this shows that $|\partial^{\boldsymbol{\gamma}}_{\mathbf{x}} K(\mathbf{x})|$ satisfies (4.4), so the kernel K satisfies the correct differential inequalities.

We also need to verify the cancellation conditions given in Definition 2.2, part (b). Let $\mathbf{x} = (\mathbf{x}', \mathbf{x}'') \in \mathbb{R}^{N_1} \times \mathbb{R}^{N_2}$ be a decomposition of the variables into two subsets, with $\mathbf{x}' = (\mathbf{x}_{p_1}, \ldots, \mathbf{x}_{p_r})$ and $\mathbf{x}'' = (\mathbf{x}_{q_1}, \ldots, \mathbf{x}_{q_t})$. Let ψ be a normalized bump function in the $\mathbf{x}''$ variables. Then

$$\begin{aligned}
&K_{\psi,R}(\mathbf{x}')\\
&= \int_{\mathbb{R}^{N_2}} K(\mathbf{x}',\mathbf{x}'')\psi(R\cdot\mathbf{x}'')\,d\mathbf{x}''\\
&= \int_{\mathbb{R}^{N}}\int_{\mathbb{R}^{N_2}} e^{2\pi i<\mathbf{x},\boldsymbol{\xi}>}m(\boldsymbol{\xi})\psi(R\cdot\mathbf{x}'')\,d\mathbf{x}''\,d\boldsymbol{\xi}\\
&= \int_{\mathbb{R}^{N_1}}\left[\int_{\mathbb{R}^{N_2}}\left[\int_{\mathbb{R}^{N_2}} e^{2\pi i<\mathbf{x}'',\boldsymbol{\xi}''>}\psi(R\cdot\mathbf{x}'')\,d\mathbf{x}''\right] m(\boldsymbol{\xi}',\boldsymbol{\xi}'')\,d\boldsymbol{\xi}''\right] e^{2\pi i<\mathbf{x}',\boldsymbol{\xi}'>}\,d\boldsymbol{\xi}'\\
&= \int_{\mathbb{R}^{N_1}}\left[\int_{\mathbb{R}^{N_2}}\left[R^{-Q''}\widehat{\psi}(R^{-1}\cdot\boldsymbol{\xi}'')\right] m(\boldsymbol{\xi}',\boldsymbol{\xi}'')\,d\boldsymbol{\xi}''\right] e^{2\pi i<\mathbf{x}',\boldsymbol{\xi}'>}\,d\boldsymbol{\xi}'\\
&= \int_{\mathbb{R}^{N_1}} e^{2\pi i<\mathbf{x}',\boldsymbol{\xi}'>}m_{\psi,R}(\boldsymbol{\xi}')\,d\boldsymbol{\xi}'
\end{aligned}$$

where

$$m_{\psi,R}(\boldsymbol{\xi}') = \int_{\mathbb{R}^{N_2}}\left[R^{-Q''}\widehat{\psi}(R^{-1}\cdot\boldsymbol{\xi}'')\right] m(\boldsymbol{\xi}',\boldsymbol{\xi}'')\,d\boldsymbol{\xi}'' = \int_{\mathbb{R}^{N_2}} \widehat{\psi}_R(\boldsymbol{\xi}'')m(\boldsymbol{\xi}',\boldsymbol{\xi}'')\,d\boldsymbol{\xi}''.$$

Since $||\widehat{\psi}_R||_{L^1}$ is finite and independent of R, it is easy to see that $m_{\psi,R}$ belongs to the class $\mathcal{M}_\infty(\widehat{N}_{p_1},\ldots,\widehat{N}_{p_r})$ uniformly on the space $\mathbb{R}^{N_1}$. Thus the estimates for $K_{\psi,R}$ follow from the same arguments. This completes the proof of Theorem 4.1. □

4.2. Fourier transforms of kernels

THEOREM 4.2. *Let $\mathcal{K}\in\mathcal{P}_0(\mathbf{E})$. Then the Fourier transform of $\mathcal{K}$ is a smooth function belonging to the class $\mathcal{M}_\infty(\mathbf{E})$.*

PROOF. Suppose that $\mathcal{K}\in\mathcal{P}_0(\mathbf{E})$. According to Definition 2.5 we can write $\mathcal{K}=\mathcal{K}_0+\psi$ where $\mathcal{K}_0\in\mathcal{P}_0(\mathbf{E})$ has compact support in $\mathbb{B}(1)$ and $\psi\in\mathcal{S}(\mathbb{R}^N)$. Since $\widehat{\psi}\in\mathcal{M}_\infty(\mathbf{E})$, it suffices to show that $\widehat{\mathcal{K}_0}\in\mathcal{M}_\infty(\mathbf{E})$. Thus without loss of generality we can assume that $\mathcal{K}$ has compact support in the ball $\mathbb{B}(1)$. The distribution $\mathcal{K}$ then acts on $\mathcal{C}^\infty(\mathbb{R}^N)$, and in particular, we set $m(\boldsymbol{\xi}) = \big\langle\mathcal{K},\exp[2\pi i<\cdot,\boldsymbol{\xi}>]\big\rangle$. From general principles, this is a smooth function and

(4.10)

$$\partial^\gamma_{\boldsymbol{\xi}} m(\boldsymbol{\xi}) = (2\pi i)^{|\gamma|}\big\langle\mathcal{K},\mathbf{x}^\gamma\exp[2\pi i<\cdot,\boldsymbol{\xi}>]\big\rangle = (2\pi i)^{|\gamma|}\big\langle\mathbf{x}^\gamma\mathcal{K},\exp[2\pi i<\cdot,\boldsymbol{\xi}>]\big\rangle$$

Note that if $|\gamma|\geq 1$, the distribution $\mathbf{x}^\gamma\mathcal{K}$ coincides with integration against the locally integrable function $\mathbf{x}^\gamma K(\mathbf{x})$, (see Remark 2.1.7 on page 36 of [**NRS01**]), and this justifies the last equality.

We want to show

$$|\partial^\gamma_{\boldsymbol{\xi}} m(\boldsymbol{\xi})| \leq C_\gamma\prod_{j=1}^n\big(1+\widehat{N}_j(\boldsymbol{\xi})\big)^{-[\![\gamma_j]\!]}. \tag{4.11}$$

Since $\mathcal{K}$ has compact support, this inequality follows if $\boldsymbol{\xi}\in\mathbb{B}(1)$. Thus we only need to consider $\boldsymbol{\xi}\in\mathbb{B}(1)^c$. Suppose $\boldsymbol{\xi}$ belongs to $\widehat{E}_S$ where $S=\big\{(I_1,k_1);\ldots;(I_s,k_s)\big\}\in$

$\mathcal{S}(n)$. Then for $j \in I_r$, $1+\widehat{N}_j(\boldsymbol{\xi}) \approx n_{k_r}(\boldsymbol{\xi}_{k_r})^{1/e(k_r,j)}$ and we want to prove
(4.12)
$$\left|\partial_{\boldsymbol{\xi}}^{\gamma} m(\boldsymbol{\xi})\right| \leq C_{\gamma} \prod_{r=1}^{s} \prod_{j \in I_r} \left(1+\widehat{N}_j(\boldsymbol{\xi})\right)^{-[\![\gamma_j]\!]} \approx C_{\gamma} \prod_{r=1}^{s} n_{k_r}(\boldsymbol{\xi}_{k_r})^{-\sum_{j \in I_r} [\![\gamma_j]\!]/e(k_r,j)}.$$

The proof now follows the same pattern as the proof of Theorem 4.1. To establish (4.12) we split the integral in (4.10) into 2^s regions depending on whether $\widehat{n}_{S,r}(\mathbf{x}_{I_r})$ is larger or smaller than $n_{k_r}(\boldsymbol{\xi}_{k_r})^{-1}$.

Let φ_0 and φ_1 be smooth functions on the positive half-line with $\varphi_0(t)+\varphi_1(t) \equiv 1$, $\varphi_0(t) \equiv 1$ for t near zero, and $\varphi_0(t) \equiv 0$ for t large. For $\epsilon = (\epsilon_1, \ldots, \epsilon_s) \in \{0,1\}^s$ set

$$\begin{aligned} K_{\epsilon}(\mathbf{x}) &= K(\mathbf{x}) \prod_{r=1}^{s} \varphi_{\epsilon_r}\big(n_{k_r}(\boldsymbol{\xi}_{k_r}) \widehat{n}_{S,r}(\mathbf{x}_{I_r})\big), \\ m_{\epsilon}(\xi) &= \int_{\mathbb{R}^N} e^{2\pi i \langle \mathbf{x}, \boldsymbol{\xi} \rangle} K_{\epsilon}(\mathbf{x})\, d\mathbf{x}. \end{aligned} \tag{4.13}$$

Note that $m(\boldsymbol{\xi}) = \sum_{\epsilon \in \{0,1\}^s} m_{\boldsymbol{\epsilon}}(\boldsymbol{\xi})$ and $K(\mathbf{x}) = \sum_{\epsilon \in \{0,1\}^s} K_{\epsilon}(\mathbf{x})$. Recall that $\widehat{n}_{S,r}(\boldsymbol{\xi}_r) = \sum_{j \in I_r} n_j(\boldsymbol{\xi}_j)^{e(k_r,j)}$ so $\widehat{n}_{S,r}$ is homogeneous relative to the family of dilations $\{\xi_j; j \in I_r\} \to \{\lambda^{1/e(k_r,j)} \xi_j : j \in I_r\}$. It follows from Proposition 15.1 that for $\boldsymbol{\delta}_{k_r} \in \mathbb{N}^{C_{k_r}}$ and $\boldsymbol{\epsilon} \in \{0,1\}^s$ we have

$$\left|\partial_{\mathbf{x}_{k_1}}^{\boldsymbol{\delta}_{k_1}} \cdots \partial_{\mathbf{x}_{k_s}}^{\boldsymbol{\delta}_{k_s}} K_{\epsilon}(\mathbf{x})\right| \leq C_{\boldsymbol{\delta}} \prod_{j=1}^{n} N_j(\mathbf{x})^{-Q_j} \prod_{r=1}^{s} \widehat{n}_{S,r}(\mathbf{x}_{k_r})^{-[\![\boldsymbol{\delta}_{k_r}]\!]} \tag{4.14}$$

since $e(k_r,k_r)=1$. Now fix $\epsilon \in \{0,1\}^s$ and let $A(\epsilon) = \{r \in \{1,\ldots,s\} : \epsilon_r = 1\}$. Let $\boldsymbol{\delta}_{k_r} \in \mathbb{N}^{c_{k_r}}$. We will choose the entries of each $\boldsymbol{\delta}_{k_r}$ to be sufficiently large integers. Then

$$\begin{aligned} &\Big(\prod_{r \in A(\boldsymbol{\epsilon})} \xi_{k_r}^{\boldsymbol{\delta}_r}\Big) \partial_{\boldsymbol{\xi}}^{\gamma} m_{\boldsymbol{\epsilon}}(\boldsymbol{\xi}) \\ &\qquad = (-2\pi i)^{|\gamma|} \int_{\mathbb{R}^N} \Big(\prod_{r \in A(\boldsymbol{\epsilon})} \xi_{k_r}^{\boldsymbol{\delta}_r}\Big) e^{-2\pi i \langle \mathbf{x}, \boldsymbol{\xi} \rangle} \mathbf{x}^{\gamma} K_{\epsilon}(\mathbf{x})\, d\mathbf{x} \\ &\qquad = (-2\pi i)^{|\gamma|-|\boldsymbol{\delta}|} \int_{\mathbb{R}^N} \Big[\prod_{r \in A(\boldsymbol{\epsilon})} \partial_{\mathbf{x}_{k_r}}^{\boldsymbol{\delta}_r}\Big] \big(e^{-2\pi i \langle \mathbf{x}, \boldsymbol{\xi} \rangle}\big)\, \mathbf{x}^{\gamma} K_{\epsilon}(\mathbf{x})\, d\mathbf{x} \\ &\qquad = (-1)^{|\boldsymbol{\delta}|} (-2\pi i)^{|\gamma|-|\boldsymbol{\delta}|} \int_{\mathbb{R}^N} e^{-2\pi i \langle \mathbf{x}, \boldsymbol{\xi} \rangle} \Big[\prod_{r \in A(\boldsymbol{\epsilon})} \partial_{\mathbf{x}_{k_r}}^{\boldsymbol{\delta}_r}\Big] (\mathbf{x}^{\gamma} K_{\epsilon})(\mathbf{x})\, d\mathbf{x}. \end{aligned} \tag{4.15}$$

The product rule shows that the derivative $\big[\prod_{r \in A(\boldsymbol{\epsilon})} \partial_{\mathbf{x}_{k_r}}^{\boldsymbol{\delta}_r}\big](\mathbf{x}^{\gamma} K_{\boldsymbol{\epsilon}})(\mathbf{x})$ can be written as a sum of terms of the form $\big(\prod_{r \in A(\boldsymbol{\epsilon})} \partial_{\mathbf{x}_{k_r}}^{\boldsymbol{\mu}_r}\big)(\mathbf{x}^{\gamma})\, \big(\prod_{r \in A(\boldsymbol{\epsilon})} \partial_{\mathbf{x}_{k_r}}^{\boldsymbol{\delta}_r - \boldsymbol{\mu}_r}(K_{\epsilon}(\mathbf{x}))\big)$. In the first factor, the entries of $\boldsymbol{\mu}_r$ can only be as large as the corresponding terms $\boldsymbol{\gamma}_r$. Thus we will choose $\boldsymbol{\delta}_r$ to have entries much larger than those of $\boldsymbol{\gamma}_r$ so that the entries of $\boldsymbol{\delta}_r - \boldsymbol{\mu}_r$ are large enough to make the integrals below converge independently of the compact support of m. Then

$$\Big(\prod_{r \in A(\boldsymbol{\epsilon})} \partial_{\mathbf{x}_{k_r}}^{\boldsymbol{\mu}_r}\Big)(\mathbf{x}^{\gamma}) = C_{\boldsymbol{\mu},\gamma} \prod_{r \in A(\epsilon)} \Big(\prod_{\substack{j \in I_r \\ j \neq k_r}} \mathbf{x}_j^{\gamma_j}\Big) \mathbf{x}_{k_r}^{\gamma_{k_r} - \boldsymbol{\mu}_{k_r}} \prod_{r \notin A(\epsilon)} \Big(\prod_{j \in I_r} \mathbf{x}_j^{\gamma_j}\Big)$$

Note that $e(j,l) \leq e(j,k_r)e(k_r,l)$. Thus

$$
\begin{aligned}
N_j(\mathbf{x}) = \sum_{l=1}^{n} |\mathbf{x}_l|^{e(j,l)} \geq \sum_{l\in I_r} |\mathbf{x}_l|^{e(j,l)} \geq \sum_{l\in I_r} |\mathbf{x}_l|^{e(k_r,l)e(J,k_r)} \\
\gtrsim \Big[\sum_{L\in I_r} |\mathbf{x}_l|^{e(k_r,l)}\Big]^{e(j,k_r)} = \widehat{n}_{S,r}(\mathbf{x}_{I_r})^{e(j,k_r)}.
\end{aligned} \tag{4.16}
$$

According to (4.14)

$$
\begin{aligned}
\Big|\Big(\prod_{r\in A(\epsilon)} \partial_{\mathbf{x}_{k_r}}^{\delta_r-\mu_r}(K_\epsilon)(\mathbf{x})\Big)\Big| &\leq C_{\delta,\mu}\prod_{j=1}^{n} N_j(\mathbf{x})^{-Q_j} \prod_{r\in A(\epsilon)} \widehat{n}_{S,r}(\mathbf{x}_{I_r})^{-[\![\delta_r]\!]+[\![\mu_r]\!]} \\
&\leq C_{\delta,\mu}\prod_{r=s} \widehat{n}_{S,r}(\mathbf{x}_{I_r})^{-\sum_{j\in I_r} Q_j/e(k_r,j)} \prod_{r\in A(\epsilon)} \widehat{n}_{S,r}(\mathbf{x}_{I_r})^{-[\![\delta_r]\!]+[\![\mu_r]\!]}.
\end{aligned}
$$

Thus according to (4.7) and (4.16), we can estimate $\big|\big(\prod_{r\in A(\epsilon)} \mathbf{x}_{k_r}^{\delta_r}\big)\partial_{\mathbf{x}}^{\gamma} K_\epsilon(\mathbf{x})\big|$ by a sum of terms of the form

$$
\begin{aligned}
&\Big[\prod_{r\notin A(\epsilon)} \int_{\substack{\mathbb{R}^{I_r}\\ \widehat{n}_{S,r}(\mathbf{x}_{I_r})<n_{k_r}(\boldsymbol{\xi}_{k_r})^{-1}}} \Big|\prod_{j\in I_r} \mathbf{x}_j^{\gamma_j}\Big| \widehat{n}_{S,r}(\mathbf{x}_{I_r})^{-\sum_{j\in I_r} Q_j e(j,k_r)}\, d\mathbf{x}_{I_r}\Big]\times \\
&\Big[\prod_{r\in A(\epsilon)} \int_{\substack{\mathbb{R}^{I_r}\\ \widehat{n}_{S,r}(\mathbf{x}_{I_r})\geq n_{k_r}(\boldsymbol{\xi}_{k_r})^{-1}}} \Big|\Big(\prod_{\substack{j\in I_r\\ j\neq k_r}} \mathbf{x}_j^{\gamma_j}\Big)\mathbf{x}_{k_r}^{\gamma_{k_r}-\mu_{k_r}}\Big| \widehat{n}_{S,r}(\mathbf{x}_{I_r})^{-\sum_{j\in I_r} Q_j e(j,k_r)-[\![\delta_r]\!]+[\![\mu_r]\!]}\, d\mathbf{x}_{I_r}\Big] \\
\approx &\Big[\prod_{r\notin A(\epsilon)} \int_{\substack{\mathbb{R}^{I_r}\\ \widehat{n}_{S,r}(\mathbf{x}_{I_r})<n_{k_r}(\boldsymbol{\xi}_{k_r})^{-1}}} \Big|\prod_{j\in I_r} \mathbf{x}_j^{\gamma_j}\Big| \widehat{n}_{S,r}(\mathbf{x}_{I_r})^{-\sum_{j\in I_r} Q_j e(j,k_r)}\, d\mathbf{x}_{I_r}\Big]\times \\
&\Big[\prod_{r\in A(\epsilon)} \int_{\substack{\mathbb{R}^{I_r}\\ \widehat{n}_{S,r}(\mathbf{x}_{I_r})\geq n_{k_r}(\boldsymbol{\xi}_{k_r})^{-1}}} \Big|\Big(\prod_{j\in I_r} \mathbf{x}_j^{\gamma_j}\Big)\Big| \widehat{n}_{S,r}(\mathbf{x}_{I_r})^{-\sum_{j\in I_r} Q_j e(j,k_r)-[\![\delta_r]\!]}\, d\mathbf{x}_{I_r}\Big].
\end{aligned}
$$

It may happen that the integrals over the regions where $\widehat{n}_{S,r}(\mathbf{x}_{I_r}) < n_{k_r}(\boldsymbol{\xi}_{k_r})^{-1}$ diverge if all the appropriate γ_j are equal to zero. However, for these terms, we use the cancellation properties of the kernels since $\varphi_{\epsilon_r}\big(n_{k_r}(\boldsymbol{\xi}_{k_r})\widehat{n}_{S,r}(\mathbf{x}_{I_r})\big)$ are dilates of normalized bump functions. Thus according to Proposition 15.3, this last expression is bounded by

$$
\begin{aligned}
&\prod_{r\notin A(\epsilon)} n_{k_r}(\boldsymbol{\xi}_{k_r})^{-\sum_{j\in I_r} [\![\gamma_j]\!]e(j,k_r)} \prod_{r\in A(\epsilon)} n_{k_r}(\boldsymbol{\xi}_{k_r})^{\sum_{j\in I_r}(-([\![\gamma_j]\!]-[\![\delta_r]\!])e(j,k_r))} \\
&\quad = \prod_{r\in A(\epsilon)} n_{k_r}(\boldsymbol{\xi}_{k_r})^{[\![\delta_r]\!]e(j,k_r)} \prod_{r=1}^{s} n_{k_r}(\boldsymbol{\xi}_{k_r})^{-\sum_{j\in I_r} [\![\gamma_j]\!]e(j,k_r)}.
\end{aligned}
$$

Comparing with (4.7), this shows that $|\partial_{\boldsymbol{\xi}}^{\gamma} m(\boldsymbol{\xi})|$ satisfies (4.12), so the multiplier m satisfies the correct differential inequalities, and this completes the proof of Theorem 4.2. $\square$

CHAPTER 5

Dyadic sums of Schwartz functions

In Chapter 4 we showed that distributions $\mathcal{K} \in \mathcal{P}_0(\mathbf{E})$ can be characterized in terms of their Fourier transforms. We now begin the study of a different kind of characterization in terms of dyadic sums of normalized bump functions. In this chapter we show that appropriate sums of dilates do converge to distributions $\mathcal{K} \in \mathcal{P}_0(\mathbf{E})$. In Theorem 6.14 below, we will see that, modulo 'coarser' kernels, every distribution $\mathcal{K} \in \mathcal{P}_0(\mathbf{E})$ can be decomposed in this way.

If $\varphi \in \mathcal{S}(\mathbb{R}^N)$, the dilates $[\varphi]_{\mathbf{t}}$ and $(\varphi)_{\mathbf{t}}$ were defined in equation (2.8). In this chapter we show that appropriate dyadic sums of dilates $[\varphi]_{\mathbf{t}}$ of Schwartz functions having suitable cancellation properties converge to distributions $\mathcal{K} \in \mathcal{P}_0(\mathbf{E})$. Similarly we show that dyadic sums of dilates $(\varphi)_{\mathbf{t}}$ of Schwartz functions with appropriate vanishing conditions converge to multipliers $\mathcal{M}_\infty(\mathbf{E})$. The main results appear in sections 5.4-5.6. However, before turning to the main results, we begin with some further remarks about cones associated to matrices, and then discuss inclusions among classes of kernels and multipliers defined by different matrices.

5.1. Cones associated to a matrix E

Let $\mathbf{E} = \{e(j,k)\}$ be an $n \times n$ matrix satisfying the basic hypotheses in (2.10) and let $\mu \geq 0$. We introduce the following notation.

$$\begin{aligned} &\Gamma(\mathbf{E}) = \Big\{\mathbf{t} = (t_1, \ldots, t_n) \in \mathbb{R}^n : \forall j \neq k,\ e(j,k)t_k \leq t_j < 0\Big\}, \\ &\Gamma_\mu(\mathbf{E}) = \Big\{\mathbf{t} = (t_1, \ldots, t_n) \in \mathbb{R}^n : \forall j \neq k,\ e(j,k)t_k - \mu \leq t_j < 0\Big\}, \\ (5.1)\quad &\Gamma^o(\mathbf{E}) = \operatorname{int}\big(\Gamma(\mathbf{E})\big) = \Big\{\mathbf{t} = (t_1, \ldots, t_n) \in \mathbb{R}^n : \forall j \neq k,\ e(j,k)t_k < t_j < 0\Big\}, \\ &\Gamma_{\mathbb{Z},\mu}(\mathbf{E}) = \big\{I = (i_1, \ldots, i_n) \in \mathbb{Z}^n : \forall j \neq k,\ e(j,k)i_k - \mu \leq i_j < 0\Big\}, \\ &\Gamma_{\mathbb{Z}}(\mathbf{E}) = \Gamma_{\mathbb{Z},0}(\mathbf{E}). \end{aligned}$$

We will refer to properties of the cone $\Gamma(\mathbf{E})$ which are proved in Appendix A. In particular, Lemma 5.1 summarizes the content of Lemma 14.7 and Lemma 14.9.

Since $\mathbf{E}$ satisfies the basic hypothesis, $\Gamma(\mathbf{E})$ is not empty.[1] However its interior, which is an open convex cone, can be empty. For example, if $\mathbf{E} = \begin{bmatrix} 1 & 2 \\ \frac{1}{2} & 1 \end{bmatrix}$, $\Gamma(\mathbf{E})$ is the one-dimensional ray given by $x_1 = 2x_2$, $x_2 < 0$, which has no interior. Note that in this case $1 = e(1,2)e(2,1)$. The following Lemma shows that, more generally, any such equality is the obstruction to the open cone being non-empty.

LEMMA 5.1. *Let* $\mathbf{E}$ *be a matrix satisfying the basic hypotheses* (2.10).

[1] See Proposition 14.5 in Appendix A.

(a) *The interior of* $\Gamma(\mathbf{E})$ *is non-empty if and only if* $\mathbf{E}$ *is reduced, i.e.,* $1 < e(j,k)e(k,j)$ *for every pair* $1 \leq j \neq k \leq n$.

(b) $\Gamma(\mathbf{E})$ *is contained in the intersection of the hyperplanes*

$$\{\mathbf{t} \in \mathbb{R}^n : e(j,k)t_k = t_j\}$$

for all pairs (j,k) *such that* $e(j,k)e(k,j) = 1$ *and has non-empty interior in this subspace of* $\mathbb{R}^n$.

5.2. Inclusions among classes of kernels associated to different matrices

The cone $\Gamma(\mathbf{E})$ contains the relevant information on the structure of kernels and multipliers associated with the matrix $\mathbf{E}$.

PROPOSITION 5.2. Let $\mathbf{E}, \mathbf{E}'$ be two $n \times n$ matrices, both satisfying the basic hypotheses. The following are equivalent:

(i) $\mathcal{P}_0(\mathbf{E}) \subseteq \mathcal{P}_0(\mathbf{E}')$;
(ii) $\mathcal{M}_\infty(\mathbf{E}) \subseteq \mathcal{M}_\infty(\mathbf{E}')$;
(iii) for every $j = 1, \dots, n$ and $\boldsymbol{\xi} \in \mathbb{B}(1)^c$, $\widehat{N}'_j(\boldsymbol{\xi}) \leq \widehat{N}_j(\boldsymbol{\xi})$;
(iv) for every $j, k = 1, \dots, n$, $e(j,k) \leq e'(j,k)$;
(v) $\Gamma(\mathbf{E}) \subseteq \Gamma(\mathbf{E}')$.

PROOF. The equivalence between (i) and (ii) is given by Theorems 4.1 and 4.2. The implications (v)⇒(iv)⇒(iii)⇒(ii) are obvious. In order to prove that (ii) implies (v), assume by contradiction that $\Gamma(\mathbf{E})$ in not contained in $\Gamma(\mathbf{E}')$. Then there exists a half-line $\mathbb{R}^-\boldsymbol{\alpha} \subset \Gamma(\mathbf{E}) \setminus \Gamma(\mathbf{E}')$, where $\boldsymbol{\alpha} = (\alpha_1, \dots, \alpha_n)$ has strictly positive components.

Let $N_{\boldsymbol{\alpha}}(\boldsymbol{\xi}) = \sum_{j=1}^n n_j(\boldsymbol{\xi}_j)^{1/\alpha_j}$. Note that $N_{\boldsymbol{\alpha}}$ is homogeneous of degree one with respect to the scaling by $\lambda^{d_j\alpha_j}$ in $\boldsymbol{\xi}_j$. Consider multipliers m which satisfy the inequalities

$$|\partial^\gamma m(\boldsymbol{\xi})| \lesssim \big(1 + N_{\boldsymbol{\alpha}}(\boldsymbol{\xi})\big)^{-\sum_j \alpha_j [\![\gamma_j]\!]} \tag{5.2}$$

for all $\boldsymbol{\gamma} \in \mathbb{N}^N$. We show that every such m is in $\mathcal{M}_\infty(\mathbf{E})$ but there exist some which are not in $\mathcal{M}_\infty(\mathbf{E}')$.

To prove the first statement it suffices to show that, for every j,

$$\big(1 + N_{\boldsymbol{\alpha}}(\boldsymbol{\xi})\big)^{\alpha_j} \gtrsim 1 + \widehat{N}_j(\boldsymbol{\xi}).$$

This condition is equivalent to the inequalities $\alpha_k/\alpha_j \leq e(k,j)$ for every j, k, and in turn these are equivalent to the condition $-\boldsymbol{\alpha} \in \Gamma(\mathbf{E})$.

Similarly, since $-\boldsymbol{\alpha} \notin \Gamma(\mathbf{E}')$, there exist j, k such that $\alpha_k/\alpha_j > e'(k,j)$. Suppose in (5.2) we take $\boldsymbol{\gamma} = (0, \dots, 0, \gamma_j, \dots, 0)$ and $\boldsymbol{\xi} = (0, \dots, 0, \boldsymbol{\xi}_k, \dots, 0)$. The inequality becomes

$$|\partial^{\gamma_j} m(0, \dots, \boldsymbol{\xi}_k, \dots, 0)| \lesssim \big(1 + n_k(\boldsymbol{\xi}_k)^{1/\alpha_k}\big)^{-\alpha_j [\![\gamma_j]\!]}$$

where the right-hand side is not dominated by

$$\big(1 + n_k(\boldsymbol{\xi}_k)\big)^{-\frac{1}{e'(k,j)}[\![\gamma_j]\!]} \approx \big(1 + \widehat{N}'_j(\boldsymbol{\xi})\big)^{-[\![\gamma_j]\!]}.$$

It is then sufficient to construct m satisfying (5.2) and such that, for some $l \in C_j$ it satisfies the opposite inequality

$$|\partial_{\xi_l} m(0, \dots, \boldsymbol{\xi}_k, \dots, 0)| \gtrsim \big(1 + n_k(\boldsymbol{\xi}_k)^{1/\alpha_k}\big)^{-\alpha_j d_l},$$

e.g.,

$$m(\boldsymbol{\xi}) = \frac{\xi_l}{\widetilde{N}_{\boldsymbol{\alpha}}(\boldsymbol{\xi})^{\alpha_j d_l}},$$

where $\widetilde{N}_{\boldsymbol{\alpha}}$ is a smooth homogeneous norm equivalent to $N_{\boldsymbol{\alpha}}$ and even in ξ_l. This m is not in $\mathcal{M}_\infty(\mathbf{E}')$. □

5.3. Coarser decompositions and lower dimensional matrices

We will often need to consider classes $\mathcal{P}_0(\mathbf{E})$ where $\mathbf{E}$ is a lower-dimensional (say $s \times s$) matrix and the variables $\mathbf{x}_1, \mathbf{x}_2, \dots, \mathbf{x}_n$ have correspondingly been grouped together into s blocks with given dilation exponents within each block.

Precisely, consider a coarser decompositions of $\mathbb{R}^N$,

$$\mathbb{R}^N = \mathbb{R}^{A_1} \oplus \mathbb{R}^{A_2} \oplus \cdots \oplus \mathbb{R}^{A_s},$$

where $\mathcal{A} = \{A_1, A_2, \dots, A_s\}$ is a partition of $\{1, 2, \dots, n\}$ and $\mathbb{R}^{A_r}$ is defined according to (2.3). We also assign an n-tuple $\boldsymbol{\alpha} = (\alpha_1, \alpha_2, \dots, \alpha_n)$ of positive exponents to define dilations on each $\mathbb{R}^{A_r}$, with homogeneous norm

$$n_{A_r}(\mathbf{t}_{A_r}) = \sum_{j \in A_r} n_j(\mathbf{t}_j)^{\alpha_j}, \qquad r = 1, \dots, n. \tag{5.3}$$

Given an $s \times s$ matrix $\mathbf{E} = \big(e(t,r)\big)_{1 \le t,r \le s}$ satisfying the basic hypotheses, with $s < n$, let $\mathcal{P}_0(\mathbf{E}, \mathcal{A}, \boldsymbol{\alpha})$ be the class of associated kernels and $\mathcal{M}_\infty(\mathbf{E}, \mathcal{A}, \boldsymbol{\alpha})$ the class of multipliers. The associated norms are, for $1 \le t \le s$,

$$\begin{aligned} N_t(\mathbf{x}) &= \sum_{r=1}^{s} n_{A_r}(\mathbf{x}_{A_r})^{e(t,r)} \approx \sum_{r=1}^{s} \sum_{k \in A_r} n_k(\mathbf{x}_k)^{\alpha_k e(t,r)} \\ \widehat{N}_t(\boldsymbol{\xi}) &= \sum_{r=1}^{s} n_{A_r}(\boldsymbol{\xi}_{A_r})^{\frac{1}{e(r,t)}} \approx \sum_{r=1}^{s} \sum_{k \in A_r} n_k(\boldsymbol{\xi}_k)^{\frac{\alpha_k}{e(r,t)}}. \end{aligned} \tag{5.4}$$

Define the $n \times n$ matrix $\mathbf{E}^\sharp = \big(e^\sharp(j,k)\big)$ where, for $j \in A_t$ and $k \in A_r$,

$$e^\sharp(j,k) = \frac{\alpha_k}{\alpha_j} e\big(t,r\big). \tag{5.5}$$

Then $\mathbf{E}^\sharp$ satisfies the basic assumptions and for $j \in A_t$ the associated dual norms are

$$\begin{aligned} N_j^\sharp(\mathbf{x}) &= \sum_{r=1}^{s} \sum_{k \in A_r} n_k(\mathbf{x}_k)^{\frac{\alpha_k}{\alpha_j} e(t,r)} \approx N_t(\mathbf{x})^{1/\alpha_j} \\ \widehat{N}_j^\sharp(\boldsymbol{\xi}) &= \sum_{r=1}^{s} \sum_{k \in A_r} n_k(\boldsymbol{\xi}_k)^{\frac{\alpha_k}{\alpha_j} \frac{1}{e(r,t)}} \approx \widehat{N}_t(\boldsymbol{\xi})^{1/\alpha_j}. \end{aligned}$$

LEMMA 5.3. *We have the equalities* $\mathcal{M}_\infty(\mathbf{E}, \mathcal{A}, \boldsymbol{\alpha}) = \mathcal{M}_\infty(\mathbf{E}^\sharp)$, $\mathcal{P}_0(\mathbf{E}, \mathcal{A}, \boldsymbol{\alpha}) = \mathcal{P}_0(\mathbf{E}^\sharp)$.

PROOF. Let $m \in \mathcal{M}_\infty(\mathbf{E}, \mathcal{A}, \boldsymbol{\alpha})$. Then, for every $\boldsymbol{\gamma} = (\gamma_1, \dots, \gamma_s) \in \mathbb{N}^{A_1} \times \cdots \times \mathbb{N}^{A_s}$,

$$\big|\partial_{\boldsymbol{\xi}}^{\boldsymbol{\gamma}} m(\boldsymbol{\xi})\big| \lesssim \prod_{t=1}^{s} \big(1 + \widehat{N}_t(\boldsymbol{\xi})\big)^{-[\![\gamma_t]\!]},$$

where

$$[\![\gamma_t]\!] = \sum_{j\in A_t} \frac{1}{\alpha_j} \sum_{l\in C_j} \gamma_l d_l,$$

by (2.7), taking into account the dilations on $\mathbb{R}^{A_t}$ which make (5.3) a homogeneous norm.

If $\gamma_t = (\gamma'_j)_{j\in A_t}$, γ can be written as an element of $\mathbb{N}^{C_1} \times \cdots \times \mathbb{N}^{C_n}$ as $\gamma = (\gamma'_1, \ldots, \gamma'_n)$. Then

$$\sum_{l\in C_j} \gamma_l d_l = [\![\gamma'_j]\!]$$

and therefore

$$\begin{aligned}\prod_{t=1}^{s} \big(1+\widehat{N}_t(\boldsymbol{\xi})\big)^{-[\![\gamma_t]\!]} &= \prod_{t=1}^{s} \prod_{j\in A_t} \big(1+\widehat{N}_t(\boldsymbol{\xi})\big)^{-\frac{1}{\alpha_j}[\![\gamma'_j]\!]} \\ &\approx \prod_{t=1}^{s} \prod_{j\in A_t} \big(1+\widehat{N}^\sharp_j(\boldsymbol{\xi})\big)^{-[\![\gamma'_j]\!]} = \prod_{j=1}^{n} \big(1+\widehat{N}^\sharp_j(\boldsymbol{\xi})\big)^{-[\![\gamma'_j]\!]} .\end{aligned}$$

Then $m \in \mathcal{M}_\infty(\mathbf{E})$. The opposite implication is proved in the same way. The equality $\mathcal{P}_0(\mathbf{E}) = \mathcal{P}_0(\mathbf{E}^\sharp)$ follows from Theorems 4.1 and 4.2. □

Notice that Lemma 5.3 implies that kernels belonging to $\mathcal{P}_0(\mathbf{E}, \mathcal{A}, \boldsymbol{\alpha})$ satisfy stronger cancellation properties than those required by the definition. Combining together Proposition 5.2 and Lemma 5.3 we obtain the following statement.

COROLLARY 5.4. Let $\mathcal{A}$ and $\boldsymbol{\alpha}$ be as above, and let $\mathbf{E}$ an $s \times s$ matrix satisfying the basic hypotheses. $\mathbf{E}'$ an $n \times n$ matrix also satisfying the basic hypotheses. We have the inclusion $\mathcal{P}_0(\mathbf{E}, \mathcal{A}, \boldsymbol{\alpha}) \subseteq \mathcal{P}_0(\mathbf{E}')$ if and only if the matrices $\mathbf{E}^\sharp$ and $\mathbf{E}'$ satisfy the condition

$$e^\sharp(j,k) \le e'(j,k), \qquad \forall\, j,k,\ 1 \le j,k \le n,$$

together with the other equivalent conditions in Proposition 5.2.

This result can be applied in particular to the case where an $n \times n$ matrix $\mathbf{E}$ satisfying the basic hypotheses is not reduced, i.e., there exist $j \neq k$ such that $e(j,k)e(k,j) = 1$. As shown in Appendix A (Section 14.5), in this case the j-th row of $\mathbf{E}$ equals $e(j,k)$ times the k-th row, and the same holds for the k-th and j-th column. Following the notation of Lemma 14.9, we denote by $k_1, \ldots, k_s$ representatives of the equivalence classes $A_1, \ldots, A_s$ under the relation $j \sim k \iff e(j,k)e(k,j) = 1$.[2] Then, for all $j \in A_t$,

$$N_j(\mathbf{x}) \approx N_{k_t}(\mathbf{x})^{e(j,k_t)} \quad \text{and} \quad \widehat{N}_j(\boldsymbol{\xi}) \approx \widehat{N}_{k_t}(\boldsymbol{\xi})^{e(j,k_t)}. \tag{5.6}$$

Moreover, for every $r \in \{1, \ldots, s\}$,

$$N_{k_t}(\mathbf{x}) \approx \sum_{r=1}^{s} \Big[\sum_{k\in A_r} n_k(\mathbf{x}_k)^{e(k_r,k)} \Big]^{e(k_t,k_r)}$$

[2] If $d(j,k) = \log\big(e(j,k)e(k,j)\big)$, then d satisfies all the requirements of a metric on the set $\{1, \ldots, n\}$ except that $d(j,k) = 0$ if and only if $e(j,k)e(k,j) = 1$. It follows that $\sim$ is an equivalence relation. Moreover, if $d(j,k) = 0$ then $d(l,k) = d(l,j)$ and hence $e(l,k) = d(j,k)e(l,j)$ for every l.

and

$$\widehat{N}_{k_t}(\boldsymbol{\xi}) \approx \sum_{r=1}^{s} \Big[\sum_{k \in A_r} n_k(\boldsymbol{\xi}_k)^{e(k_r,k)} \Big]^{1/e(k_r,k_t)}.$$

Therefore we are in the previous situation, with $\alpha_k = e(k_r, k)$ if $k \in A_r$ and with the reduced matrix

$$\mathbf{E}^\flat = \big(e(k_t, k_r)\big)_{1 \le t,r \le s}$$

defined in Section 14.5. Notice that $(\mathbf{E}^\flat)^\sharp = \mathbf{E}$. By Lemma 5.3 we have the following

PROPOSITION 5.5. $\mathcal{P}_0(\mathbf{E}) = \mathcal{P}_0(\mathbf{E}^\flat)$ and $\mathcal{M}_\infty(\mathbf{E}) = \mathcal{M}_\infty(\mathbf{E}^\flat)$.

Proposition 5.5 makes it possible to replace, whenever it is appropriate, $\mathbf{E}$ by $\mathbf{E}^\flat$ and reduce matters to the case of a reduced matrix.

5.4. Size estimates

We now turn to the construction of the dyadic decompositions of kernels K belonging to classes $\mathcal{P}_0(\mathbf{E})$. In the rest of this section, we shall assume that $\Gamma^o(\mathbf{E})$ is non-empty, and consequently that $\Gamma_\mathbb{Z}(\mathbf{E})$ is non-empty, where $\Gamma^o(\mathbf{E})$ and $\Gamma_\mathbb{Z}(\mathbf{E})$ were defined in equation (5.1). Let $\{\varphi^I : I \in \Gamma_\mathbb{Z}(\mathbf{E})\}$ be a uniformly bounded family in $\mathcal{C}_0^\infty(\mathbb{R}^N)$ or $\mathcal{S}(\mathbb{R}^N)$. We will show that if all the functions φ^I satisfy a suitable cancellation condition, then the sum

$$K = \sum_{I \in \Gamma_\mathbb{Z}(\mathbf{E})} [\varphi^I]_I \tag{5.7}$$

converges in the sense of distributions to an element $\mathcal{K} \in \mathcal{P}_0(\mathbf{E})$. The precise statement is given in Theorem 5.13 below.

We first show that, even without any cancellation hypotheses on the functions φ^I, the sum in equation (5.7) satisfies the differential inequalities of Definition 2.2 for the class $\mathcal{P}_0(\mathbf{E})$.

LEMMA 5.6. *Suppose that $\Gamma^o(\mathbf{E})$ is non-empty, and let $\{\varphi^I : I \in \Gamma_\mathbb{Z}(\mathbf{E})\} \subset \mathcal{S}(\mathbb{R}^N)$ be a uniformly bounded family of Schwartz functions. Then the series $K(\mathbf{x}) = \sum_{I\in\Gamma_\mathbb{Z}(\mathbf{E})}[\varphi^I]_I(\mathbf{x})$ converges absolutely for all $\mathbf{x} \neq 0$, $K \in \mathcal{C}^\infty(\mathbb{R}^N \setminus \{0\})$, and*

$$\partial^\gamma K(\mathbf{x}) = \sum_{I \in \Gamma_\mathbb{Z}(\mathbf{E})} 2^{-\sum_{j=1}^n i_j(Q_j + [\![\gamma_j]\!])} (\partial^\gamma \varphi^I)\big(2^{-i_1} \cdot \mathbf{x}_1, \dots, 2^{-i_n} \cdot \mathbf{x}_n\big). \tag{5.8}$$

Moreover, for every $\gamma = (\gamma_1, \dots, \gamma_n) \in \mathbb{N}^N$ and $M \geq 0$ there exists a constant $C_{\gamma,M} > 0$ so that

$$\big|\partial^\gamma K(\mathbf{x})\big| \le C_{\gamma,M} \prod_{j=1}^n N_j(\mathbf{x})^{-(Q_j+[\![\gamma_j]\!])}(1+|\mathbf{x}|)^{-M}. \tag{5.9}$$

PROOF. Since the family $\{\varphi^I\}$ is uniformly bounded in $\mathcal{S}(\mathbb{R}^n)$, for each $M > 0$ there is a constant $C_{\gamma,M}$ so that

$$\Big|2^{-\sum_{j=1}^n i_j(Q_j+[\![\gamma_j]\!])}(\partial^\gamma\varphi^I)\big(2^{-i_1}\cdot\mathbf{x}_1,\dots,2^{-i_n}\cdot\mathbf{x}_n\big)\Big| \\ \le C_{\gamma,M}\prod_{j=1}^n 2^{-i_j(Q_j+[\![\gamma_j]\!])}\Big(1+\sum_{j=1}^n 2^{-i_j} n_j(\mathbf{x}_j)\Big)^{-M}.$$

It follows from Proposition 16.3 in Appendix C that the series on the right hand side of equation (5.8) converges absolutely and is bounded by

$$C\prod_{j=1}^{n} N_j(\mathbf{x})^{-(Q_j+[\![\gamma_j]\!])}\prod_{j=1}^{n}(1+n_j(\mathbf{x}))^{-M'}.$$

This shows that the formal differentiation of equation (5.7) yielding (5.8) is justified, and also establishes the inequality in (5.9). □

As an application of Lemma 5.6 we check that kernels satisfying the differential inequalities in part (a) of Definition 2.2 just fail to be integrable near the origin. See also Lemma 7.6.

COROLLARY 5.7. Let $\mathbf{E}$ satisfy (2.10) and let $N_j(\mathbf{x})$ be defined in equation (2.9). Suppose that $\Gamma^o(\mathbf{E})$ is non-empty. Then

(a) $\displaystyle\int_{\mathbb{B}(1)}\prod_{j=1}^{n} N_j(\mathbf{x})^{-Q_j}\,d\mathbf{x} = +\infty;$

(b) If $\alpha > 0$ and $1 \leq k \leq n$ then $\displaystyle\int_{\mathbb{B}(1)} N_k(\mathbf{x})^{\alpha}\prod_{j=1}^{n} N_j(\mathbf{x})^{-Q_j}\,d\mathbf{x} < +\infty$.

PROOF. Choose $\varphi \in \mathcal{C}_0^\infty(\mathbb{R}^N)$ supported in $\mathbb{B}(1)$ with $\int_{\mathbb{R}^N}\varphi(\mathbf{x})\,d\mathbf{x} = 1$ and $\varphi(\mathbf{x}) \geq 0$. Then if $\varphi^I = \varphi$ for every I, the family $\{\varphi^I\}$ is normalized in $\mathcal{S}(\mathbb{R}^N)$. According to equation (5.9), if $F \subset \Gamma(\mathbf{E})$ is a finite set with M elements , then $\sum_{I\in F}[\varphi]_I(\mathbf{x}) \leq C\prod_{j=1}^N N_j(\mathbf{x})^{-Q_j}$, and hence

$$\int_{\mathbb{B}(1)}\prod_{j=1}^{N} N_j(\mathbf{x})^{-Q_j}\,d\mathbf{x} \geq C^{-1}\sum_{I\in F}\int_{\mathbb{R}^N}[\varphi]_I(\mathbf{x})\,d\mathbf{x} = C^{-1}M.$$

Letting $M\to\infty$ gives statement (a). To prove statement (b), observe that there is a constant $c_{k,l} > 0$ so that if $\mathbf{x}\in\mathbb{B}(1)$, then $N_k(\mathbf{x}) \leq N_l(\mathbf{x})^{c_{k,l}}$. Thus given $\alpha > 0$ and $1 \leq k \leq n$ there exist positive numbers $\epsilon_1,\ldots,\epsilon_n$ so that $N_k(\mathbf{x})^\alpha \leq \prod_{l=1}^n N_l(\mathbf{x})^{\epsilon_l}$. Then

$$\begin{aligned}\int_{\mathbb{B}(1)} N_k(\mathbf{x})^{\alpha}\prod_{j=1}^{n} N_j(\mathbf{x})^{-Q_j}\,d\mathbf{x} &\leq \int_{\mathbb{B}(1)}\prod_{j=1}^{n} N_j(\mathbf{x})^{-Q_j+\epsilon_j}\,d\mathbf{x}\\ &\leq \prod_{j=1}^{n}\int_{n_j(\mathbf{x}_j)\leq 1} n_j(\mathbf{x}_j)^{-Q_j+\epsilon_j}\,d\mathbf{x}_j < +\infty.\end{aligned}$$

This completes the proof. □

5.5. Cancellation properties

We next turn to a discussion of cancellation properties. Let $\mathbb{R}^N = \bigoplus_{j=1}^n \mathbb{R}^{C_j}$ and suppose $\mathbf{E}$ is an $n\times n$ matrix satisfying (2.10).

DEFINITION 5.8. Let $\varphi \in \mathcal{S}(\mathbb{R}^N)$.

(a) φ has cancellation in the variable $\mathbf{x}_j$ if $\displaystyle\int_{\mathbb{R}^{C_j}}\varphi(\mathbf{x}_1,\ldots,\mathbf{x}_j,\ldots,\mathbf{x}_n)\,d\mathbf{x}_j = 0.$

(b) φ has strong cancellations if it has cancellation in $\mathbf{x}_j$ for $1 \leq j \leq n$.

This terminology was used in Definition 5.3 of [**NRSW12**]. We will also need a variant of the notion of *weak-cancellation* that was introduced in Definition 5.5 of [**NRSW12**].

DEFINITION 5.9. A function $\varphi \in \mathcal{C}_0^\infty(\mathbb{R}^N)$ has weak cancellation with parameter $\epsilon > 0$ relative to the n-tuple $I = (i_1, \ldots, i_n) \in \Gamma_{\mathbb{Z}}(\mathbf{E})$ provided that one can write

$$\varphi = \sum_{A \subset \{1,\ldots,n-1\}} \Big[\prod_{j \in A} 2^{-\epsilon\Lambda(I,j)}\Big]\varphi_A$$

where the sum is taken over all subsets $A \subset \{1, \ldots, n-1\}$, and for each subset A,

(1) $\varphi_A \in \mathcal{C}_0^\infty(\mathbb{R}^N)$, and is normalized relative to φ;
(2) φ_A has cancellation in the variable $\mathbf{x}_j$ for each $j \notin A$, and in particular for $j = n$;[3]
(3) $\tau_A : A \to \{1, \ldots, n\}$ is a mapping with $\tau_A(j) \neq j$ for all $j \in A$;
(4) $\Lambda(I, j) = i_j - e\big(j, \tau_A(j)\big)\, i_{\tau_A(j)}$.

Note that if $I \in \Gamma_{\mathbb{Z}}(\mathbf{E})$, then $i_j - e\big(j, \tau_A(j)\big)\, i_{\tau_A(j)} \geq 0$. Thus if φ has weak cancellation, it can be written as a finite sum of functions, each of which has the property that for each index j, either there is cancellation in $\mathbf{x}_j$ or the lack of cancellation is compensated by a gain $2^{-\epsilon[i_j - e(j,\tau_A(j))\, i_{\tau_A(j)}]}$.

The following is Lemma 5.1 in [**NRSW12**]:

PROPOSITION 5.10. If $\varphi \in \mathcal{S}(\mathbb{R}^N)$ and $\{J_1, \ldots, J_r\}$ are disjoint subsets of $\{1, \ldots, N\}$, then the following are equivalent:

(a) $\int \varphi(\mathbf{x})\, d\mathbf{x}_{J_k} = 0$ for $1 \leq k \leq r$.
(b) For every $(j_1, \ldots, j_r) \in J_1 \times \cdots \times J_r$ there is a function $\varphi_{j_1,\ldots,j_r} \in \mathcal{S}(\mathbb{R}^n)$, normalized with respect to φ, so that

$$\varphi(\mathbf{x}) = \sum_{j_1 \in J_1} \cdots \sum_{j_r \in J_r} \partial_{j_1} \cdots \partial_{j_r} \varphi_{j_1,\ldots,j_r}(\mathbf{x}).$$

Moreover, if $\varphi \in \mathcal{C}_0^\infty(\mathbb{R}^N)$, then we can choose the functions $\varphi_{j_1,\ldots,j_r} \in \mathcal{C}_0^\infty(\mathbb{R}^N)$ normalized relative to φ.

We will also need the following result.

PROPOSITION 5.11. Let $L = \{l_1, \ldots, l_a\}$ and $M = \{m_1, \ldots, m_b\}$ be complementary subsets of $\{1, \ldots, n\}$, let A be a subset of M, and let $R = \{R_{m_1}, \ldots, R_{m_b}\}$ be positive real numbers. Write any $\mathbf{x} \in \mathbb{R}^N$ as $\mathbf{x} = (\mathbf{x}', \mathbf{x}'')$ where

$$\mathbf{x}' = (\mathbf{x}_{l_1}, \ldots, \mathbf{x}_{l_a}) \in \mathbb{R}^L \quad \text{and} \quad \mathbf{x}'' = (\mathbf{x}_{m_1}, \ldots, \mathbf{x}_{m_b}) \in \mathbb{R}^M.$$

Let $\varphi \in \mathcal{S}(\mathbb{R}^N)$ and let $\psi \in \mathcal{C}_0^\infty(\mathbb{R}^M)$ have support in $\mathbb{B}(1)$. Put

$$\Phi(\mathbf{x}') = \int_{\mathbb{R}^M} \varphi(\mathbf{x}', \mathbf{x}'')\psi(R_{m_1} \cdot \mathbf{x}_{m_1}, \ldots, R_{m_b} \cdot \mathbf{x}_{m_b})\, d\mathbf{x}_{m_1} \cdots d\mathbf{x}_{m_b}.$$

(a) $\Phi \in \mathcal{S}\left(\mathbb{R}^L\right)$ and is normalized relative to φ and ψ with constants independent of R. If $\varphi \in \mathcal{C}_0^\infty(\mathbb{R}^N)$, then $\Phi \in \mathcal{C}_0^\infty\left(\mathbb{R}^L\right)$ and is normalized relative to φ and ψ with constants independent of R.

[3]The special role of $\mathbf{x}_n$ reflects our later consideration of convolution on nilpotent groups where these variables belong to the center of the group.

(b) If some of the parameters $\{R_{m_j}\}$ are large, there is an improvement over statement (a): there exists $\epsilon > 0$ independent of φ, ψ and R so that

$$\Phi(\mathbf{x}') = \Big[\prod_{m_j \in M} \min\{1, R_{m_j}^{-\epsilon}\}\Big]\,\widetilde{\Phi}(\mathbf{x}')$$

where $\widetilde{\Phi} \in \mathcal{S}(\mathbb{R}^N)$ is normalized relative to φ and ψ, with constants independent of R. If $\varphi \in \mathcal{C}_0^\infty(\mathbb{R}^N)$ is supported in the unit ball, then $\widetilde{\Phi} \in \mathcal{C}_0^\infty(\mathbb{R}^L)$ and is normalized relative to φ and ψ with constants independent of R.

(c) If φ has cancellation in some of the variables in M there is an improvement over statement (b) even if R_{m_j} is small. There exists $\epsilon > 0$ independent of φ, ψ and R so that if φ has cancellation in $\mathbf{x}_j$ for each $j \in A$ then

$$\Phi(\mathbf{x}') = \Big[\prod_{m_j \in M\setminus A} \min\{1, R_{m_j}^{-\epsilon}\}\Big]\Big[\prod_{m_j \in A} \min\{R_{m_j}^{\epsilon}, R_{m_j}^{-\epsilon}\}\Big]\,\widetilde{\Phi}(\mathbf{x}'),$$

where $\widetilde{\Phi}$ is normalized relative to φ and ψ, with constants independent of of R.

PROOF. Statement (a) is clear. Since $\psi(\mathbf{x}'')$ is supported where $n_{m_j}(\mathbf{x}_{m_j}) \leq 1$, we have $n_{m_j}(\mathbf{x}_{m_j}) \leq R_{m_j}^{-1}$ if $\psi(R_{m_1}\cdot\mathbf{x}_{m_1}, \ldots, R_{m_b}\cdot\mathbf{x}_{m_b}) \neq 0$. The integration in the definition of $\Phi(\mathbf{x}')$ takes place over the set $\{(\mathbf{x}_{m_1}, \ldots, \mathbf{x}_{m_b}) : n_{m_j}(\mathbf{x}_{m_j}) \leq R_{m_j}^{-1}\}$, and we can estimate the integral by the (small) size of the region of integration in the variables $\{\mathbf{x}_{m_j}\}$ for which R_{m_j} is large. This gives the estimate in part (b). To establish (c) we use Proposition 5.10: $\Phi(\mathbf{x}')$ is a sum of terms of the form

$$\int_{\mathbb{R}^M} \partial_{j_1}\cdots\partial_{j_a}\varphi_{j_1,\ldots,j_a}(\mathbf{x}', \mathbf{x}'')\psi(R_{m_1}\cdot\mathbf{x}_{m_1}, \ldots, R_{m_b}\cdot\mathbf{x}_{m_b})\,d\mathbf{x}_{m_1}\cdots d\mathbf{x}_{m_b}$$

where ∂_{j_k} is a derivative with respect to a variable in $\mathbf{x}_{m_k}$ for each $m_k \in A$. Integrating by parts in the variables in A for which $R_{m_k} < 1$ then gives the estimate in part (c), and this completes the proof. □

We will want to show that sums of dilates of normalized bump functions with weak cancellation satisfy the cancellation properties of the distributions $\mathcal{K} \in \mathcal{P}_0(\mathbf{E})$, and we will use the following.

LEMMA 5.12. *Suppose that φ^I has weak cancellation with parameter $\epsilon > 0$. Let $L = \{l_1, \ldots, l_a\}$ and $M = \{m_1, \ldots, m_b\}$ be complementary subsets of $\{1, \ldots, n\}$, and let $R = \{R_1, \ldots, R_a\}$ be positive real numbers. Write $\mathbf{x} \in \mathbb{R}^N$ as $\mathbf{x} = (\mathbf{x}', \mathbf{x}'')$ where $\mathbf{x}' = (\mathbf{x}_{l_1}, \ldots, \mathbf{x}_{l_a})$ and $\mathbf{x}'' = (\mathbf{x}_{m_1}, \ldots, \mathbf{x}_{m_b})$, and similarly write $I = (I', I'')$. Let $\psi \in \mathcal{C}_0^\infty(\mathbb{R}^L)$ be a normalized bump function. Then the function*

$$\Phi^I_{\psi,R}(\mathbf{x}'') = \int_{\mathbb{R}^L} \varphi^I(\mathbf{x}', \mathbf{x}'')\psi\big((2^{i_{l_1}}R_1)\cdot\mathbf{x}_{l_1}, \ldots, (2^{i_{l_a}}R_a)\cdot\mathbf{x}_{l_a}\big)\,d\mathbf{x}_{l_1}\cdots d\mathbf{x}_{l_a}.$$

can be written as a finite sum of terms, indexed by subsets $A \subset \{1, \ldots, n-1\}$, of the form

$$\Big[\prod_{j \in A\cap L} 2^{-\epsilon[i_j - e(j,\tau_A(j))i_{\tau_A(j)}]}\Big]\Big[\prod_{j \in L\setminus A} \min\Big\{(2^{i_{l_j}}R_j)^{+\epsilon}, (2^{i_{l_j}}R_j)^{-\epsilon}\Big\}\Big]\,\widetilde{\varphi}^I_A(\mathbf{x}''),$$

where τ_A is the mapping introduced in Definition 5.9. The family $\{\widetilde{\varphi}^I_A\} \subset \mathcal{C}_0^\infty(\mathbb{R}^L)$ is normalized relative to φ^I.

PROOF. According to Definition 5.9, since φ^I has weak cancellation,

$$\varphi^I(\mathbf{x}', \mathbf{x}'') = \sum_{A \subset \{1,\dots,n-1\}} \Big[\prod_{j \in A} 2^{-\epsilon[i_j - e(j,\tau_A(j)) i_{\tau_A(j)}]}\Big] \varphi^I_A(\mathbf{x}', \mathbf{x}'')$$

where φ^I_A has cancellation in $\mathbf{x}_j$ for each $j \notin A$. Thus $\Phi^I_{\psi,R}$ can be written as a finite sum of terms, indexed by $A \subset \{1, \dots, n-1\}$, of the form

$$\Big[\prod_{j \in A} 2^{-\epsilon[i_j - e(j,\tau_A(j)) i_{\tau_A(j)}]}\Big] \int \varphi^I_A(\mathbf{x}', \mathbf{x}'')\psi\big((2^{i_{l_1}} R_1) \cdot \mathbf{x}_{l_1}, \dots, (2^{i_{l_a}} R_a) \cdot \mathbf{x}_{l_a}\big)\, d\mathbf{x}'.$$

Using Proposition 5.11, and incorporating $\prod_{j \in L \setminus A} 2^{-\epsilon[i_j - e(j,\tau_A(j)) i_{\tau_A(j)}]}$ into the bump function, it follows this function can be written

$$\Big[\prod_{j \in A \cap M} 2^{-\epsilon[i_j - e(j,\tau_A(j)) i_{\tau_A(j)}]}\Big]\Big[\prod_{j \in M \setminus A} \min\Big\{(2^{i_j} R_{l_j})^{+\epsilon}, (2^{i_{l_j}} R_j)^{-\epsilon}\Big\}\Big] \widetilde{\varphi}^I_A(\mathbf{x}''),$$

which completes the proof. □

5.6. Dyadic sums with weak cancellation

Fix the decomposition $\mathbb{R}^N = \mathbb{R}^{C_1} \oplus \cdots \oplus \mathbb{R}^{C_n}$ and let $\mathbf{E}$ be an $n \times n$ matrix satisfying the basic hypothesis given in equation (2.10). Suppose that $\Gamma^o(\mathbf{E}) \neq \emptyset$. Let $\Gamma_{\mathbb{Z}}(\mathbf{E})$ be as in equation (5.1).

THEOREM 5.13. *Let $\{\varphi^I : I \in \Gamma_{\mathbb{Z}}(\mathbf{E})\} \subset \mathcal{S}(\mathbb{R}^N)$ be a uniformly bounded family. If each function φ^I has weak cancellation with parameter $\epsilon > 0$ relative to I, then $K = \sum_{I \in \Gamma_{\mathbb{Z}}(\mathbf{E})} [\varphi^I]_I$ converges in the sense of distributions to a kernel $\mathcal{K} \in \mathcal{P}_0(\mathbf{E})$.*

The arguments are similar to those in the proofs of Theorem 6.8 and Propositions 6.9, 6.10, and 6.11 of [**NRSW12**]. The proof consists of two steps:

(1) If $F \subset \Gamma_{\mathbb{Z}}(\mathbf{E})$ is a finite set, the distribution $\mathcal{K}_F = \sum_{I \in F} [\varphi^I]_I$ belongs to the class $\mathcal{P}_0(\mathbf{E})$ with constants independent of the set F. (Lemma 5.6 shows that $\mathcal{K}_F$ satisfies the appropriate differential inequalities, so we only need to verify that $\mathcal{K}_F$ satisfies the cancellation conditions.)
(2) If $\{F_m\}$ is any increasing sequence of finite subsets of $\Gamma_{\mathbb{Z}}(\mathbf{E})$ with $\bigcup F_m = \Gamma_{\mathbb{Z}}(\mathbf{E})$, then $\lim_{m \to \infty} \mathcal{K}_{F_m}$ exists in the sense of distributions.

PROOF OF STEP 1.

Let $F \subset \Gamma_{\mathbb{Z}}(\mathbf{E})$ be a finite set and let $\mathcal{K}_F = \sum_{I \in F} [\varphi^I]_I$. To verify the cancellation conditions, let $L = \{l_1, \dots, l_a\}$ and $M = \{m_1, \dots, m_b\}$ be complementary subsets of $\{1, \dots, n\}$. If $\mathbf{x} = (\mathbf{x}_1, \dots, \mathbf{x}_n) \in \mathbb{R}^N$ write $\mathbf{x} = (\mathbf{x}', \mathbf{x}'')$ where $\mathbf{x}' = (\mathbf{x}_{l_1}, \dots, \mathbf{x}_{l_a}) \in \mathbb{R}^L$ and $\mathbf{x}'' = (\mathbf{x}_{m_1}, \dots, \mathbf{x}_{m_b}) \in \mathbb{R}^M$. Let $\psi \in \mathcal{C}^\infty_0(\mathbb{R}^L)$ be a normalized bump function with support in the unit ball, and let $R = \{R_{l_1}, \dots, R_{l_a}\}$ be positive real numbers. We must show that the function

$$K_{F,\psi,R}(\mathbf{x}'') = \sum_{I \in F} \int_{\mathbb{R}^M} [\varphi^I]_I(\mathbf{x}_1, \dots, \mathbf{x}_n)\psi(R_{l_1} \cdot \mathbf{x}_{l_1}, \dots, R_{l_a} \cdot \mathbf{x}_{l_a})\, d\mathbf{x}_{l_1} \cdots d\mathbf{x}_{l_a}$$

satisfies the correct differential inequalities in the $\mathbf{x}''$ variables.

To simplify the notation, and without any real loss of generality, we will only consider the case when $L = \{1, \dots, a\}$ and $M = \{a+1, \dots, n\}$, so that $\mathbf{x}' =$

$(\mathbf{x}_1, \ldots, \mathbf{x}_a)$ and $\mathbf{x}'' = (\mathbf{x}_{a+1}, \ldots, \mathbf{x}_n)$. Let $\mathbf{E}''$ be the sub-matrix of $\mathbf{E}$ consisting of those entries $e(j,k)$ for which $a+1 \leq j,k \leq n$, and let

$$\Gamma_{\mathbb{Z}}(\mathbf{E}'') = \Big\{(i_{a+1}, \ldots, i_n) \in \mathbb{Z}^{n-a} : e(j,k)i_k \leq i_j < 0,\, a+1 \leq j,k \leq n\Big\}.$$

For each $I'' = (i_{a+1}, \ldots, i_n) \in \Gamma_{\mathbb{Z}}(\mathbf{E}'')$ there are only finitely many elements $I \in \Gamma_{\mathbb{Z}}(\mathbf{E})$ which have the same last $n-a$ entries. Let $F'' \subset \Gamma_{\mathbb{Z}}(\mathbf{E}'')$ be a finite set and suppose (without loss of generality) that

$$F = \Big\{I = (i_1, \ldots, i_n) \in \Gamma_{\mathbb{Z}}(\mathbf{E}) : I'' = (i_{a+1}, \ldots, i_n) \in F''\Big\}.$$

For each $I'' \in \Gamma_{\mathbb{Z}}(\mathbf{E}'')$ let

$$\Lambda(I'') = \Big\{I' = (i_1, \ldots, i_a) \in \mathbb{Z}^a : (I', I'') \in \Gamma_{\mathbb{Z}}(\mathbf{E})\Big\}.$$

Then for each $I \in F$ we can write $I = (I', I'')$ with $I'' \in F''$ and $I' \in \Lambda(I'')$.

Let $\psi \in \mathcal{C}_0^\infty(\mathbb{R}^M)$ be a normalized bump function with support in the unit ball, and let $R = \{R_1, \ldots, R_a\}$ be positive real numbers. To complete step 1 we need to estimate

$$\begin{aligned}
&K_{F,\psi,R}(\mathbf{x}'') \\
&\quad = \sum_{I'' \in F''} \sum_{I' \in \Lambda(I'')} \int_{\mathbb{R}^M} [\varphi^I]_I(\mathbf{x}_1, \ldots, \mathbf{x}_n)\psi(R_1 \cdot \mathbf{x}_1, \ldots, R_a \cdot \mathbf{x}_a)\, d\mathbf{x}_1 \cdots d\mathbf{x}_a
\end{aligned}$$

and its derivatives. Let

$$\Phi^I_{\psi,R}(\mathbf{x}'') = \int_{\mathbb{R}^M} \varphi^I(\mathbf{x}_1, \ldots, \mathbf{x}_n)\psi\big((2^{i_1}R_1) \cdot \mathbf{x}_1, \ldots, (2^{i_a}R_a) \cdot \mathbf{x}_a\big)\, d\mathbf{x}_1 \cdots d\mathbf{x}_a.$$

Then

$$\int_{\mathbb{R}^M} [\varphi^I]_I(\mathbf{x}_1, \ldots, \mathbf{x}_n)\psi(R_1 \cdot \mathbf{x}_1, \ldots, R_a \cdot \mathbf{x}_a)\, d\mathbf{x}_1 \cdots d\mathbf{x}_a = \big[\Phi^{(I',I'')}_{\psi,R}\big]_{I''}(\mathbf{x}'')$$

and so

$$K_{F,\psi,R}(\tilde{\mathbf{x}}'') = \sum_{I'' \in F'} \Big[\sum_{I' \in \Lambda(I'')} \Phi^{(I',I'')}_{\psi,R}\Big]_{I''}(\mathbf{x}'').$$

If we can show that for each $I'' \in F''$ the inner sum $\sum_{I' \in \Lambda(I'')} \Phi^{(I',I'')}_{\psi,R}$ converges to a function $\widetilde{\Phi}^{I''}_{\psi,R} \in \mathcal{C}_0^\infty(\mathbb{R}^L)$, and this family is uniformly bounded independent of the choice of F, then the required differential inequality estimates for $K_{F,\psi,R}$ will follow from Lemma 5.6. But according to Lemma 5.12, $\Phi^{(I',I'')}_{\psi,R}$ is a sum of normalized bump functions multiplied by exponential gains. Thus the proof of step 1 will be complete if we can show that for each fixed $I'' \in F'' \subset \Gamma_{\mathbb{Z}}(\mathbf{E}'')$ and each subset $A \subset \{1, \ldots, n\}$,

$$\sum_{(i_1,\ldots,i_a) \in \Lambda(I'')} \Bigg[\prod_{\substack{j=1 \\ j \in A \cap L}}^{a} 2^{-\epsilon[i_j - e(j,\tau_A(j))i_{\tau_A(j)}]}\Bigg]\Bigg[\prod_{\substack{j=1 \\ j \in L \setminus A}}^{a} \min\Big\{(2^{i_j}R_j)^{+\epsilon}, (2^{i_j}R_j)^{-\epsilon}\Big\}\Bigg]$$

converges independently of I'' and the finite set F''. Without loss of generality, we can assume that $\{1, \ldots, a\} \cap A = \{1, \ldots, b\}$ and $\{1, \ldots, a\} \setminus A = \{b+1, \ldots, a\}$. Thus we need to estimate

$$\sum_{(i_1,\ldots,i_a) \in \Lambda(I'')} \Bigg[\prod_{j=1}^{b} 2^{-\epsilon[i_j - e(j,\tau_A(j))i_{\tau_A(j)}]}\Bigg]\Bigg[\prod_{j=b+1}^{a} \min\Big\{(2^{i_j}R_j)^{+\epsilon}, (2^{i_j}R_j)^{-\epsilon}\Big\}\Bigg].$$

The required estimate follows from Proposition 16.2 in Appendix C. □

PROOF OF STEP 2.

We need to check that if $\{F_m \subset \Gamma_{\mathbb{Z}}(\mathbf{E})\}$ is a sequence of finite sets with $\bigcup F_m = \Gamma_{\mathbb{Z}}(\mathbf{E})$, then the sequence $\{\mathcal{K}_{F_m} = \sum_{I \in F_m} [\varphi^I]_I\}$ converges in the sense of distributions to an element of $\mathcal{P}_0(\mathbf{E})$. The argument is similar to that on pages 667-668 of [**NRSW12**]. Since each φ^I has cancellation in $\mathbf{x}_n$, it follows from Proposition 5.11 that we can write $\varphi^I = \sum_{k \in C_n} \partial_{x_k} \varphi_k^I$ where each $\varphi_k^I \in \mathcal{C}_0^\infty(\mathbb{R}^N)$ is normalized relative to φ^I. It follows that

$$[\varphi^I]_I = \sum_{k \in C_n} [\partial_{x_k} \varphi_k^I]_I = \sum_{k \in C_n} 2^{i_n d_k} \partial_{x_k} [\varphi_k^I]_I.$$

Let $d = \min\{d_k : 1 \leq k \leq N\} > 0$ so that (since $i_n < 0$) we have $2^{i_n d_k} \leq 2^{i_n d}$. Then if $\psi \in \mathcal{S}_0^\infty(\mathbb{R}^N)$ is a test function, we can integrate by parts to get

$$\begin{aligned}
&\langle \mathcal{K}_{F_m}, \psi \rangle \\
&\quad = \sum_{k \in C_n} \Big\langle \sum_{I \in F_m} 2^{i_n d_k} \partial_{x_k} [\varphi_k^I]_I, \psi \Big\rangle \\
&\quad = -\sum_{k \in C_n} \Big\langle \sum_{I \in F_m} 2^{i_n d_k} [\varphi_k^I]_I, \partial_{x_k} \psi \Big\rangle \\
&\quad = -\sum_{k \in C_n} \int_{\mathbb{R}^N} \sum_{I \in F_m} \Big[2^{i_n d_k} 2^{-\sum_{j=1}^n i_j Q_j} \varphi_k^I (2^{-i_1} \cdot \mathbf{x}_1, \dots, 2^{-i_n} \cdot \mathbf{x}_n) \Big] \partial_{x_k} \psi(\mathbf{x})\, d\mathbf{x} \\
&\quad = -\sum_{k \in C_n} \int_{\mathbb{R}^N} \sum_{I \in F_m} \Big(\prod_{j=1}^n 2^{-i_j Q_j^\#} \Big) \varphi_k^I (2^{-i_1} \cdot \mathbf{x}_1, \dots, 2^{-i_n} \cdot \mathbf{x}_n)\, \partial_{x_k} \psi(\mathbf{x})\, d\mathbf{x}
\end{aligned}$$

where

$$Q_j^\# = \begin{cases} Q_j & \text{if } 1 \leq j \leq n-1 \\ Q_n - d & \text{if } j = n \end{cases}.$$

Since $\{\varphi_k^I\}$ is a normalized family in $\mathcal{S}(\mathbb{R}^N)$, for any $M > 0$ there is a constant C_M so that

$$\begin{aligned}
&\Big| \sum_{I \in F_m} \Big(\prod_{j=1}^n 2^{-i_j Q_j^\#} \Big) \varphi_k^I (2^{-i_1} \cdot \mathbf{x}_1, \dots, 2^{-i_n} \cdot \mathbf{x}_n) \Big| \\
&\qquad \leq C_M \sum_{I \subset \Gamma(\mathbf{E})} \Big[\Big(\prod_{j-1}^n 2^{-i_j Q_j^\#} \Big) \Big] \Big(1 + \sum_{k=1}^n 2^{-i_k} n_k(\mathbf{x}_k) \Big)^{-M}.
\end{aligned}$$

If M is sufficiently large, we can apply Proposition 16.3 with $\alpha_j = Q_j^\#$, $A_k = n_k(\mathbf{x}_k)$, and $M_j = 0$. It follows that there is a constant C so that

$$\begin{aligned}
\Big| \sum_{I \in F_m} \Big(\prod_{j=1}^n 2^{-i_j Q_j^\#} \Big) \varphi_k^I (2^{-i_1} \cdot \mathbf{x}_1, \dots, 2^{-i_n} \cdot \mathbf{x}_n) \Big| &\leq C \prod_{j=1}^n \Big[\sum_{k=1}^n n_k(\mathbf{x}_k)^{e(j,k)} \Big]^{-Q_j^\#} \\
&= C\, N_n(\mathbf{x})^d \prod_{j=1}^n N_j(\mathbf{x})^{-Q_j}.
\end{aligned}$$

It follows from Corollary 5.7 that $N_n(\mathbf{x})^{\mathbf{d}} \prod_{j=1}^n N_j(\mathbf{x})^{-Q_j}$ is locally integrable. Thus by the Lebesgue dominated convergence theorem it follows that as m tends to infinity, $\langle \mathcal{K}_{F_m}, \psi \rangle$ tends to

$$-\sum_{k \in C_n} \int_{\mathbb{R}^N} \Big[\sum_{I \in F_m} 2^{i_k d_k} 2^{-\sum_{k=1}^{n} i_k Q_k} \varphi_k^I (2^{-i_1} \cdot \mathbf{x}_1, \ldots, 2^{-i_n} \cdot \mathbf{x}_n) \Big] \partial_{x_k} \psi(\mathbf{x}) d\mathbf{x}.$$

This completes the proof. □

CHAPTER 6

Decomposition of multipliers and kernels

Let $\mathcal{K} \in \mathcal{P}_0(\mathbf{E})$. The object of this chapter is to decompose $\mathcal{K}$ into a finite sum of distributions, each of which can be written as a dyadic sum of dilates of normalized Schwartz functions (or normalized bump functions) with appropriate cancellation. It is difficult to directly decompose $\mathcal{K}$ and maintain the cancellation, so we work on the Fourier transform side. The effect of this procedure is that the following steps will be governed by the decomposition of $\mathbb{B}(1)^c$ in the $\boldsymbol{\xi}$-space according to the sets $\widehat{E}_S$, while the decomposition of $\mathbb{B}(1)$ in the $\mathbf{x}$-space will no longer play any rôle.

6.1. New matrices $\mathbf{E}_S$

Let $\mathcal{S}'(\mathbf{E})$ denote the set of marked partitions S such that $\widehat{E}_S \cap \mathbb{B}(1)^c$ is non-empty. We have the following equivalent characterization of this property.

PROPOSITION 6.1. Let $S = \big((I_1, k_1); \dots ; (I_s, k_s)\big) \in \mathcal{S}(n)$. Then $\widehat{E}_S \cap \mathbb{B}(1)^c \neq \emptyset$ if and only if the cone

$$\widehat{\Gamma}_S = \Big\{\mathbf{a} = (a_1, \dots, a_s) \in \mathbb{R}^s : \tau_S(k_r, k_p)a_p \leq a_r < 0 \text{ for all } 1 \leq p, r \leq s\Big\},$$

has non-empty interior.

PROOF. Recall from Remark 3.2 that if $\mathbf{t} \in \mathbb{B}(1)^c$ then $n_{k_r}(\mathbf{t}_r) > 1$ for all r. Now according to Corollary 3.10, if $\mathbf{t} = (\mathbf{t}_1, \dots, \mathbf{t}_n) \in \widehat{E}_S \cap \mathbb{B}(1)^c$, then

$$\tau_S(k_p, k_r) \log\big[n_{k_r}(\mathbf{t}_{k_r})^{-1}\big] < \log\big[n_{k_p}(\mathbf{t}_{k_p})^{-1}\big] < 0 \tag{6.1}$$

and so the point $\mathbf{a} = \big(\log\big[n_{k_1}(\mathbf{t}_{k_1})^{-1}\big], \dots, \log\big[n_{k_s}(\mathbf{t}_{k_s})^{-1}\big]\big) \in \widehat{\Gamma}_S$. Moreover, it follows from (6.1) that $\tau_S(k_p, k_r)\tau_S(k_r, k_p) > 1$ and so by Lemma 5.1, the interior of the cone is non-empty.

Conversely, if the interior of $\widehat{\Gamma}_S$ is non-empty, choose a point $\mathbf{a} = (a_1, \dots, a_s)$ belonging to the interior of the cone $\widehat{\Gamma}_S$, choose a point $\mathbf{t}_{k_r} \in \mathbb{R}^{C_{k_r}}$ with $n_{k_r}(\mathbf{t}_{k_r}) = e^{-a_r} > 1$. Then $n_{k_r}(\mathbf{t}_{k_r}) > n_{k_p}(\mathbf{t}_{k_p})^{1/\tau_S(k_p,k_r)}$ for all $1 \leq p, r \leq s$ with $p \neq r$. We can then choose points $\mathbf{t}_j \in \mathbb{R}^{C_j}$ for each $j \in I_r$ with $j \neq r$ such that $n_j(\mathbf{t}_j)^{1/\tau_S(j,k_r)} < n_{k_r}(\mathbf{t}_{k_r})$, and it follows from Corollary 3.10 that $\mathbf{t} = (\mathbf{t}_1, \dots, \mathbf{t}_n) \in \widehat{E}_S \cap \mathbb{B}(1)^c$. □

With the decomposition introduced in Section 3.5 and in the notation of Section 5.3, for each $S \in \mathcal{S}'(\mathbf{E})$ we want to define new norms on $\mathbb{R}^N$ which describe the behaviour of multipliers on the set $\widehat{E}_S \cap \mathbb{B}(1)^c$ in the same way that the original norms $\widehat{N}_1(\mathbf{t}), \dots, \widehat{N}_n(\mathbf{t})$ describe it on the principal region $\widehat{E}_{S_0} \cap \mathbb{B}(1)^c$.

Let $S = \{(I_1, k_1); \ldots; (I_s, k_s)\} \in \mathcal{S}'(\mathbf{E})$. On each $\mathbb{R}^{I_r}$ we put the norm

$$\widehat{n}_{S,r}(\boldsymbol{\xi}_{I_r}) = \sum_{j \in I_r} n_j(\boldsymbol{\xi}_j)^{e(k_r,j)} = \sum_{j \in I_r} n_j(\boldsymbol{\xi}_j)^{1/\tau_S(j,k_r)},$$

defined in (3.5) and homogeneous with respect to the dilations $\hat{\cdot}_S$.

We now want an $s \times s$ matrix $\mathbf{E}_S$ satisfying the basic hypotheses and such that $\Gamma(\mathbf{E}_S) = \widehat{\Gamma}_S$. We cannot simply use the coefficients $\tau_S(k_r, k_p)$ because in general they do not satisfy the inequalities $\tau_S(k_r, k_p) \leq \tau_S(k_r, k_u)\, \tau_S(k_u, k_p)$.

Luckily we can replace the coefficients $\tau_S(k_p, k_r)$ by new ones which do satisfy the basic hypothesis. According to Lemma 14.2 in Appendix A, if the cone is non-empty, the coefficients in the inequalities defining the cone can be replaced by new coefficients which satisfy the basic hypothesis. Thus we have the following result.

LEMMA 6.2. *Let $S = \{(I_1, k_1); \ldots, (I_s, k_s)\} \in \mathcal{S}'(\mathbf{E})$. There is a unique $s \times s$ matrix $\mathbf{E}_S = \{e_S(r,p)\}$ such that $e_S(r,p) \leq \tau_S(k_r, k_p)$, and*

$$\Gamma(\mathbf{E}_S) = \left\{\mathbf{a} \in \mathbb{R}^s : e_S(r,p) a_p \leq a_r < 0\right\} = \widehat{\Gamma}_S.$$

According to (5.4), we introduce global norms $N_{S,r}$ and $\widehat{N}_{S,r}$.

DEFINITION 6.3. Let $S = \{(I_1, k_1); \ldots, (I_s, k_s)\} \in \mathcal{S}'(\mathbf{E})$. Set

$$(6.2)\qquad \begin{aligned} N_{S,r}(\mathbf{x}) &= \sum_{p=1}^{s} \widehat{n}_{S,p}(\mathbf{x}_{I_p})^{e_S(r,p)} \approx \sum_{p=1}^{s} \sum_{k \in I_p} n_k(\mathbf{x}_k)^{e(k_p,k) e_S(r,p)}, \\ \widehat{N}_{S,r}(\boldsymbol{\xi}) &= \sum_{p=1}^{s} \widehat{n}_{S,p}(\boldsymbol{\xi}_{I_p})^{1/e_S(p,r)} \approx \sum_{p=1}^{s} \sum_{k \in I_p} n_k(\boldsymbol{\xi}_k)^{e(k_p,k)/e_S(p,r)}. \end{aligned}$$

These norms coincide with the norms defined in (5.4), relative to the partition $\mathcal{I}_S = \{I_1, \ldots, I_s\}$ of $\{1, \ldots, n\}$, to the matrix $\mathbf{E}_S$ and to the vector of exponents $\boldsymbol{\alpha}_S = (\alpha_1, \ldots, \alpha_n)$ with $\alpha_j = e(k_r, j)$ if $j \in I_r$.

We will use the shortened notation $\mathcal{M}_\infty(\mathbf{E}_S)$ instead of $\mathcal{M}_\infty(\mathbf{E}_S, \mathcal{I}_S, \boldsymbol{\alpha}_{\mathbb{S}})$.

PROPOSITION 6.4. We have the inclusion $\mathcal{M}_\infty(\mathbf{E}_S) \subset \mathcal{M}_\infty(\mathbf{E})$.

PROOF. By Corollary 5.4, it is sufficient to check that the entries of the matrix $\mathbf{E}_S^\sharp$ are not greater than the corresponding entries of the matrix $\mathbf{E}$. By (5.5), if $j \in I_r$ and $k \in I_p$,

$$e_S^\sharp(j,k) = \frac{e(k_p,k)}{e(k_r,j)} e_S(r,p) \leq \frac{e(k_p,k)}{e(k_r,j)} \tau_S(k_r,k_p) \leq \frac{e(k_p,k)}{e(k_r,j)} \frac{e(k_r,k)}{e(k_p,k)} \leq e(j,k).$$

□

6.2. Road map for the dyadic decomposition

Let $m = \widehat{\mathcal{K}}$ be the Fourier transform so that $m \in \mathcal{M}_\infty(\mathbf{E})$. We proceed as follows.

(1) In Section 6.3 we will construct a partition of unity on $\mathbb{R}^N$ consisting of functions Ψ_0 and $\{\Psi_S\}_{S \in \mathcal{S}'(\mathbf{E})}$, where $\Psi_0 \in \mathcal{C}_0^\infty(\mathbb{R}^N)$ has support in $\mathbb{B}(2)$, and where each $\Psi_S \in \mathcal{M}_\infty(\mathbf{E})$ and has support in $\widehat{E}_S^A \cap \mathbb{B}(1)^c$. Here $A > 1$ depends only on the matrix $\mathbf{E}$.

We will write

$$m = \Psi_0 m + \sum_{S\in\mathcal{S}'(\mathbf{E})} \Psi_S m = m_0 + \sum_{S\in\mathcal{S}'(\mathbf{E})} m_S.$$

Then $m_0 \in \mathcal{C}_0^\infty(\mathbb{B}(2))$ and each $m_S \in \mathcal{M}_\infty(\mathbf{E})$ is supported in $\widehat{E}_S^A \cap \mathbb{B}(1)^c$.

(2) For each $S = \{(I_1, k_1); \dots; (I_s, k_s)\} \in \mathcal{S}'(\mathbf{E})$, we show that

$$m_S(\boldsymbol{\xi}_{I_1}, \dots, \boldsymbol{\xi}_{I_s}) = \sum_{J\in\Gamma(\mathbf{E}_S)} m_S^J(2^{-j_1} \hat{\cdot}_S \boldsymbol{\xi}_{I_1}, \dots, 2^{-j_s} \hat{\cdot}_S \boldsymbol{\xi}_{I_s})$$

where $\{m_S^J : J \in \Gamma(\mathbf{E}_S)\} \subset \mathcal{C}_0^\infty(\mathbb{R}^N)$ is a uniformly bounded family with supports in the set where $4^{-1} < \widehat{n}_{S,r}(\boldsymbol{\xi}_{I_r}) < 4$ for $1 \le r \le s$.

(3) Writing $[m_S^J]^J(\boldsymbol{\xi}_{I_1}, \dots, \boldsymbol{\xi}_{I_s}) = m_S^J(2^{-j_1} \hat{\cdot}_S \boldsymbol{\xi}_{I_1}, \dots, 2^{-j_s} \hat{\cdot}_S \boldsymbol{\xi}_{I_s})$, it then follows that

$$m(\boldsymbol{\xi}) = m_0(\boldsymbol{\xi}) + \sum_{S\in\mathcal{S}'([n])} \sum_{J\in\Gamma(\mathbf{E}_S)} [m_S^J]^J(\boldsymbol{\xi}).$$

This is done in Section 6.4.

(4) Taking inverse Fourier transform, denoted by $^\vee$, we obtain

$$\mathcal{K} = m_0^\vee + \sum_{S\in\mathcal{S}'([n])} \sum_{J\in\Gamma(\mathbf{E}_S)} \big([m_S^J]^J\big)^\vee.$$

This gives a decomposition of $\mathcal{K}$ into sums of dilates of Schwartz functions, which in turn can be written as sums of dilates of normalized bump functions. (See Lemma 6.15 below.)

6.3. Partitions of unity

In this section, we construct a partition of unity on the 'Fourier transform side'. We begin by slightly modifying our notation. As usual, $\mathbb{R}^N = \bigoplus_{j=1}^n \mathbb{R}^{C_j}$ and $\mathbf{E} = \{e(j,k)\}$ is an $n \times n$ matrix satisfying (2.10). Let $S = \{(I_1, k_1); \dots; (I_s, k_s)\} \in \mathcal{S}'(\mathbf{E})$ and $\mathbb{R}^{I_r} = \bigoplus_{j\in I_r} \mathbb{R}^{C_j}$. Instead of taking $\widehat{N}_j(\boldsymbol{\xi}) = \sum_{l=1}^n n_l(\boldsymbol{\xi}_l)^{1/e(l,j)}$ which might not be differentiable, we choose a smooth version of this norm. Then

$$\widehat{N}_j(\boldsymbol{\xi}) \approx \sum_{l=1}^n n_l(\boldsymbol{\xi}_l)^{1/e(l,j)}, \qquad \widehat{n}_{S,r}(\boldsymbol{\xi}_{I_r}) \approx \sum_{j\in I_r} n_j(\boldsymbol{\xi}_j)^{e(k_r,j)},$$

$$\lambda \hat{\cdot}_S \boldsymbol{\xi}_{I_r} = \{\lambda^{1/e(k_r,j)} \cdot \boldsymbol{\zeta}_j : j \subset I_r\}, \qquad \widehat{Q}_{S,r} = \sum_{j\in I_r} Q_j e(k_r, j)^{-1},$$

$$\tau_S(k_p, k_r) = \min_{j\in I_r} \{e(k_p, j)/e(k_r, j)\}.$$

Let $\widehat{E}_S^A$ be the set introduced in Definition 3.5 for $A \ge 1$. Recall from Lemma 3.9 that there is a constant $\eta > 1$ so that

$$\boldsymbol{\xi} \in \widehat{E}_S^A \cap \mathbb{B}(1)^c \Longrightarrow \begin{cases} n_j(\boldsymbol{\xi}_j)^{e(k_r,j)} < A n_{k_r}(\boldsymbol{\xi}_{k_r}) & \text{for all } j \ne k_p \text{ in } I_r \\ n_{k_p}(\boldsymbol{\xi}_{k_p}) < \big(A n_{k_r}(\boldsymbol{\xi}_{k_r})\big)^{\tau_S(k_p,k_r)} & \text{for all } 1 \le p \ne r \le s \end{cases}$$

$$\Longrightarrow \boldsymbol{\xi} \in \widehat{E}_S^{A^\eta} \cap \mathbb{B}(1)^c.$$

Now pick $\psi \in \mathcal{C}^\infty(\mathbb{R})$ and $\chi \in \mathcal{C}^\infty(\mathbb{R}^N)$ such that

$$\psi(t) = \begin{cases} 1 & \text{if } t \leq 2n \\ 0 & \text{if } t \geq 4n \end{cases}, \quad \text{and} \quad \chi(\boldsymbol{\xi}) = \begin{cases} 0 & \text{if } |\boldsymbol{\xi}| \leq 1 \\ 1 & \text{if } |\boldsymbol{\xi}| \geq \frac{3}{2} \end{cases}.$$

For $S = \{(I_1, k_1); \dots; (I_s, k_s)\} \in \mathcal{S}'(\mathbf{E})$, set

$$\Psi_S^\#(\boldsymbol{\xi}) = \chi(\boldsymbol{\xi}) \prod_{r=1}^{s} \prod_{j \in I_r} \psi\big(\widehat{N}_j(\boldsymbol{\xi}) n_{k_r}(\boldsymbol{\xi}_{k_r})^{-1/e(k_r,j)}\big). \tag{6.3}$$

PROPOSITION 6.5.

(a) Let $\overline{\widehat{E}_S}$ be the closure of $\widehat{E}_S$. Then $\Psi_S^\sharp(\boldsymbol{\xi}) \equiv 1$ in an open neighborhood of the set $\overline{\widehat{E}_S} \setminus \mathbb{B}(2)$.

(b) There is a constant $A > 1$ so that $\Psi_S^\sharp$ is supported in $\widehat{E}_S^A \setminus \overline{\mathbb{B}(1)}$.

(c) $\Psi_S^\sharp \in \mathcal{C}^\infty(\mathbb{R}^N)$.

PROOF. Suppose that $\boldsymbol{\xi} \in \overline{\widehat{E}_S} \setminus \mathbb{B}(2)$. Then $\chi(\boldsymbol{\eta}) = 1$ for $\boldsymbol{\eta}$ in a neighborhood of $\boldsymbol{\xi}$. Moreover, $n_l(\boldsymbol{\xi}_l)^{1/e(l,j)} \leq n_{k_r}(\boldsymbol{\xi}_{k_r})^{1/e(k_r,j)}$ for every $j \in I_r$, and so

$$\widehat{N}_j(\boldsymbol{\xi}) n_{k_r}(\boldsymbol{\xi}_{k_r})^{-1/e(k_r,j)} = n_{k_r}(\boldsymbol{\xi}_{k_r})^{-1/e(k_r,j)} \sum_{l=1}^{n} n_l(\boldsymbol{\xi}_l)^{1/e(l,j)} \leq n.$$

It follows that $\psi\big(\widehat{N}_j(\boldsymbol{\eta}) n_{k_r}(\boldsymbol{\eta}_{k_r})^{-1/e(k_r,j)}\big) = 1$ for $\boldsymbol{\eta}$ in an open neighborhood of $\boldsymbol{\xi}$, and so the same is true for $\Psi_S^\sharp(\boldsymbol{\eta})$.

Next suppose $\Psi_S^\#(\boldsymbol{\xi}) \neq 0$. Then $\chi(\boldsymbol{\xi}) \neq 0$, and $\psi\big(\widehat{N}_j(\boldsymbol{\xi}) n_{k_r}(\boldsymbol{\xi}_{k_r})^{-1/e(k_r,j)}\big) \neq 0$ for $1 \leq r \leq s$ and every $j \in I_r$. This shows that $\boldsymbol{\xi} \in \overline{B(1)}^c$, and since $n_j(\boldsymbol{\xi}) \lesssim \widehat{N}_j(\boldsymbol{\xi}_j)$ and $n_{k_p}(\boldsymbol{\xi}_{k_p}) \lesssim \widehat{N}_j(\boldsymbol{\xi})^{e(k_p,j)}$, it follows from the assumption on the support of ψ that

$$\begin{aligned} n_j(\boldsymbol{\xi}_j) &\lesssim n_{k_r}(\boldsymbol{\xi}_{k_r})^{1/e(k_r,j)} && \text{for } j \in I_r, \\ n_{k_p}(\boldsymbol{\xi}_{k_p}) &\lesssim n_{k_r}(\boldsymbol{\xi}_{k_r})^{e(k_p,j)/e(k_r,j)} && \text{for } 1 \leq p, r \leq s. \end{aligned}$$

It follows that there exists a constant $A > 0$ (depending on n and the matrix $\mathbf{E}$) so that $n_j(\boldsymbol{\xi}_j) \leq A\, n_{k_r}(\boldsymbol{\xi}_{k_r})^{1/e(k_r,j)}$ and $n_{k_p}(\boldsymbol{\xi}_{k_p}) \leq A\, n_{k_r}(\boldsymbol{\xi}_{k_r})^{\tau_S(k_p,k_r)}$. But this means that the support of $\Psi_S^\sharp$ is contained in the set $\widehat{E}_S^{A^\eta}$ where η is the constant from Lemma 3.9.

Finally, if $\boldsymbol{\xi} \notin \mathbb{B}(1)$ and if $\Psi_S^\sharp(\boldsymbol{\xi}) \neq 0$, then $\widehat{N}_j(\boldsymbol{\xi}) \geq 1$ for $1 \leq j \leq n$ and it follows that $n_{k_r}(\boldsymbol{\xi}_{k_r})$ is bounded away from 0. The function $\widehat{N}_j(\boldsymbol{\xi}) n_{k_r}(\boldsymbol{\xi}_{k_r})^{-1/e(k_r,j)}$ is smooth on the set where $\boldsymbol{\xi}_{k_r} \neq \mathbf{0}$, and it follows that $\Psi_S^\#$ is smooth. □

LEMMA 6.6. $\Psi_S^\# \in \mathcal{M}_\infty(\widehat{E})$; *i.e. for every* $\boldsymbol{\gamma} \in \mathbb{N}^{C_1} \times \cdots \times \mathbb{N}^{C_n}$ *there is a constant* $C_{\boldsymbol{\gamma}} > 0$ *so that*

$$|\partial_{\boldsymbol{\xi}}^{\boldsymbol{\gamma}} \Psi_S^\#(\boldsymbol{\xi})| \leq C_{\boldsymbol{\gamma}} \prod_{j=1}^{n} \Big[1 + \widehat{N}_j(\boldsymbol{\xi})\Big]^{-[\![\gamma_j]\!]}.$$

PROOF. The function $\widehat{N}_j(\boldsymbol{\xi})$ is homogeneous of degree one relative to the dilations

$$(\boldsymbol{\xi}_1, \dots, \boldsymbol{\xi}_n) \to (\lambda^{e(1,j)} \cdot \boldsymbol{\xi}_1, \dots, \lambda^{e(n,j)} \cdot \boldsymbol{\xi}_n).$$

and the function $\rho_{j,k_r}(\boldsymbol{\xi}) = \widehat{N}_j(\boldsymbol{\xi}) n_{k_r}(\boldsymbol{\xi}_{k_r})^{-1/e(k_r,j)}$ is homogeneous of degree zero. The homogeneous dimension of $\mathbb{R}^{C_l}$ is $e(l,j)Q_l$, and if $\boldsymbol{\gamma} = (\gamma_1, \ldots, \gamma_n) \in \mathbb{N}^{C_1} \times \cdots \times \mathbb{N}^{C_n}$, the adapted length of $\boldsymbol{\gamma}$ is $\sum_{l=1}^{n} e(l,j)[\![\gamma_j]\!]$. It follows from Proposition 15.1 in Appendix B that for $\boldsymbol{\xi} \in \mathbb{B}(1)^c$

$$|\partial^{\gamma} \rho_{j,k_r}(\boldsymbol{\xi})| \lesssim \Big[1 + \widehat{N}_j(\boldsymbol{\xi})\Big]^{-\sum_{l=1}^{n} e(l,j)[\![\gamma_l]\!]}.$$

But on the support of $\Psi_S^{\#}$ which is contained in $\widehat{E}_S^A$,

$$\begin{aligned}
&\Big[1+\widehat{N}_j(\boldsymbol{\xi})\Big]^{-\sum_{l=1}^{n} e(l,j)[\![\gamma_l]\!]} \\
&\quad = \prod_{p=1}^{s} \prod_{l \in I_p} \Big[1 + \widehat{N}_j(\boldsymbol{\xi})\Big]^{-[\![\gamma_l]\!] e(l,j)} \leq \prod_{p=1}^{s} \prod_{l \in I_p} \Big[1 + n_{k_p}(\boldsymbol{\xi}_{k_p})\Big]^{-[\![\gamma_l]\!] e(l,j)/e(k_p,j)} \\
&\quad \leq \prod_{p=1}^{s} \prod_{l \in I_p} \Big[1 + n_{k_p}(\boldsymbol{\xi}_{k_p})\Big]^{-[\![\gamma_l]\!]/e(k_p,l)} \lesssim \prod_{p=1}^{s} \prod_{l \in I_p} \Big[1 + N_l(\boldsymbol{\xi})\Big]^{-[\![\gamma_l]\!]}
\end{aligned}$$

since $e(k_p,j) \leq e(k_p,l)e(l,j)$, $n_{k_p}(\boldsymbol{\xi}_{k_p}) > 1$, and $N_l(\boldsymbol{\xi}) \approx n_{k_p}(\boldsymbol{\xi}_{k_p})^{1/e(k_p,l)}$ on the support of $\Psi_S^{\#}$ if $l \in I_p$. The estimates of the Lemma now follow from the chain rule and product rule. □

Let $\Psi_0^{\#}$ be a compactly supported smooth function which is identically equal to 1 on $\mathbb{B}(2)$. We set

$$\Psi_0 = \Psi_0^{\#}\Big[\Psi_0^{\#} + \sum_{S \in \mathcal{S}'(\mathbf{E})} \Psi_S^{\#}\Big]^{-1}, \qquad \Psi_S = \Psi_S^{\#}\Big[\Psi_0^{\#} + \sum_{S \in \mathcal{S}'(\mathbf{E})} \Psi_S^{\#}\Big]^{-1}. \tag{6.4}$$

Using the fact that, if $m \in \mathcal{M}_\infty(\mathbf{E})$ is bounded away from zero, then also $1/m \in \mathcal{M}_\infty(\mathbf{E})$, we then have

COROLLARY 6.7.

(a) $\Psi_0(\boldsymbol{\xi}) + \sum_{S \in \mathcal{S}'(\mathbf{E})} \Psi_S(\boldsymbol{\xi}) \equiv 1$ for every $\boldsymbol{\xi} \in \mathbb{R}^N$.

(b) $\Psi_S \in \mathcal{M}_\infty(\mathbf{E})$ for $S = 0$ and for every $S \in \mathcal{S}'(\mathbf{E})$.

(c) For $S \in \mathcal{S}'(\mathbf{E})$ the function Ψ_S is supported in $\widehat{E}_S^A$.

We now use Corollary 6.7 to obtain a decomposition of a multiplier $m \in \mathcal{M}_\infty(\mathbf{E})$.

COROLLARY 6.8. Let $m \in \mathcal{M}_\infty(\mathbf{E})$. Put $m_0 = \Psi_0 m$, and $m_S = \Psi_S m$ for each $S \in \mathcal{S}'(\mathbf{E})$. Then

(a) $m(\boldsymbol{\xi}) = m_0(\boldsymbol{\xi}) + \sum_{S \in \mathcal{S}'(\mathbf{E})} m_S(\boldsymbol{\xi})$ for every $\boldsymbol{\xi} \in \mathbb{R}^N$;

(b) $m_S \in \mathcal{M}_\infty(\widehat{E})$ for every $S \in \mathcal{S}'(\mathbf{E})$;

(c) m_S is supported in $\widehat{E}_S^A$ where A depends only on the matrix $\mathbf{E}$;

(d) $m_0 \in \mathcal{C}_0^\infty(\mathbb{R}^N)$ has support in $\mathbb{B}(2)$.

REMARK 6.9. Since m_S is supported on $\widehat{E}_S^A$, we have

$$\begin{aligned}|\partial_{\boldsymbol{\xi}}^{\gamma}(m_S)(\boldsymbol{\xi})| &\lesssim \prod_{r=1}^{s}\prod_{j\in I_r}\widehat{N}_j(\boldsymbol{\xi})^{-[\![\gamma_j]\!]} \approx \prod_{r=1}^{s}\prod_{j\in I_r} n_{k_r}(\boldsymbol{\xi}_{k_r})^{-[\![\gamma_j]\!]/e(k_r,j)}\\ &\approx \prod_{r=1}^{s}\widehat{n}_{S,r}(\boldsymbol{\xi}_{I_r})^{-\sum_{j\in I_r}[\![\gamma_j]\!]/e(k_r,j)}.\end{aligned}$$

This says that the differential estimates for m_S are equivalent to product multiplier estimates relative to the norms $\{\widehat{n}_{S,r}\}$ on the decomposition $\mathbb{R}^N = \mathbb{R}^{I_1}\oplus\cdots\oplus\mathbb{R}^{I_s}$.

All the functions m_S belongs to the class $\mathcal{M}_\infty(\mathbf{E})$. If

$$S = \{(I_1,k_1);\ldots;(I_s,k_s)\} \in \mathcal{S}'(\mathbf{E})$$

is not the principal marked partition, one can say more: the function m_S in fact belongs to the class of multipliers $\mathcal{M}_\infty(\mathbf{E}_S)$ introduced in Section 6.1.

PROPOSITION 6.10. *For $S \in \mathcal{S}'(\mathbf{E})$, the function m_S belongs to the class $\mathcal{M}_\infty(\mathbf{E}_S)$.*

PROOF. We must show that for any $\gamma = (\gamma_{I_1},\ldots,\gamma_{I_s}) \in \mathbb{N}^{I_1}\times\cdots\times\mathbb{N}^{I_s}$ there is a constant C_γ so that

$$|\partial^\gamma m_S(\boldsymbol{\xi})| \leq C_\gamma \prod_{r=1}^{s}\left(1+\widehat{N}_{S,r}(\boldsymbol{\xi})\right)^{-[\![\gamma_{I_r}]\!]_S},$$

where $[\![\gamma_{I_r}]\!]_S = \sum_{j\in I_r}[\![\gamma_j]\!]/e(k_r,j)$ is the length of the multi-index γ_{I_r} adapted to the dilations $\hat{\cdot}_S$. But this is a simple consequence of the estimate in Remark 6.9 and the fact that on the set $\widehat{E}_S^A$ we have $\widehat{n}_{S,r}(\boldsymbol{\xi}_{I_r}) \approx 1+\widehat{N}_{S,r}(\boldsymbol{\xi})$. □

6.4. Dyadic decomposition of a multiplier

We now turn to the task of providing a dyadic decomposition for each multiplier m_S. Fix $S = \{(I_1,k_1);\ldots;(I_s,k_s)\} \in \mathcal{S}'(\mathbf{E})$, and write $\mathbb{R}^N = \mathbb{R}^{I_1}\oplus\cdots\oplus\mathbb{R}^{I_s}$. On each factor $\mathbb{R}^{I_r}$ choose a function $\theta \in \mathcal{C}_0^\infty(\mathbb{R}^{I_r})$ satisfying[1]

$$\theta(\boldsymbol{\xi}_{I_r}) = \begin{cases}1 & \text{if } \widehat{n}_{S,r}(\boldsymbol{\xi}_{I_r}) \leq 1,\\ 0 & \text{if } \widehat{n}_{S,r}(\boldsymbol{\xi}_{I_r}) \geq 2.\end{cases}$$

Recall that $2^{-j}\hat{\cdot}_S\,\boldsymbol{\xi}_{I_r} = \{2^{-j/e(k_r,j)}\cdot\boldsymbol{\xi}_j\}_{j\in I_r}$. Put $\theta_j(\boldsymbol{\xi}_{I_r}) = \theta(2^{-j}\hat{\cdot}_S\,\boldsymbol{\xi}_{I_r})$, and

$$\Theta_j(\boldsymbol{\xi}_{I_r}) = \theta_j(\boldsymbol{\xi}_{I_r}) - \theta_{j-1}(\boldsymbol{\xi}_{I_r}) = \theta(2^{-j}\hat{\cdot}_S\,\boldsymbol{\xi}_{I_r}) - \theta(2^{j-1}\hat{\cdot}_S\,\boldsymbol{\xi}_{I_r}) = \Theta_0(2^{-j}\hat{\cdot}_S\,\boldsymbol{\xi}_{I_r}).$$

Note that Θ_j is supported where $\widehat{n}_{S,r}(\boldsymbol{\xi}_{I_r}) \approx 2^j$. Precisely:

$$\Theta_j(\boldsymbol{\xi}_{I_r}) \neq 0 \quad\Longrightarrow\quad 2^{j-1} \leq \widehat{n}_{S,r}(\boldsymbol{\xi}_{I_r}) \leq 2^{j+2}.$$

For $0 \neq \boldsymbol{\xi}_{I_r} \in \mathbb{R}^{I_r}$ we have

$$\sum_{j=-\infty}^{\infty}\Theta_j(\boldsymbol{\xi}_{I_r}) = \theta_0(\boldsymbol{\xi}_{I_r}) + \sum_{j=1}^{\infty}\Theta_j(\boldsymbol{\xi}_{I_r}) = 1,$$

and the second equality holds for $\boldsymbol{\xi}_{I_r} = 0$ as well. For $J = (j_1,\ldots,j_s) \in \mathbb{Z}^s$ define

$$\Theta_J(\boldsymbol{\xi}) = \Theta_J(\boldsymbol{\xi}_{I_1},\ldots,\boldsymbol{\xi}_{I_s}) = \Theta_{j_1}(\boldsymbol{\xi}_{I_1})\cdots\Theta_{j_s}(\boldsymbol{\xi}_{I_s}).$$

---

[1] To keep notation simpler, we do not indicate the dependence of θ on the index r.

Then Θ_J is supported where each $\widehat{n}_{S,r}(\boldsymbol{\xi}_{I_r}) \approx 2^{j_r}$, and $\sum_{J\in\mathbb{Z}^s}\Theta_J(\boldsymbol{\xi}_{I_1},\ldots,\boldsymbol{\xi}_{I_s}) \equiv 1$ provided each $\boldsymbol{\xi}_{I_r} \neq 0$. Recall that m_S is supported on $\widehat{E}_S^A \cap \mathbb{B}(1)^c$, and $\boldsymbol{\xi}_{I_r} \neq 0$ on this support. Thus we can write

$$m_S(\boldsymbol{\xi}) = \sum_{J\in\mathbb{Z}^s}\Theta_J(\boldsymbol{\xi})m_S(\boldsymbol{\xi}) = \sum_{J\in\mathbb{Z}^s}\Theta_J(\boldsymbol{\xi})\Psi_S(\boldsymbol{\xi})m(\boldsymbol{\xi}).$$

PROPOSITION 6.11. There is a constant $\mu > 0$ depending only on $\mathbf{E}$ so that if $J \in \mathbb{Z}^s$ and $\Theta_J(\boldsymbol{\xi})\Psi_S(\boldsymbol{\xi})m(\boldsymbol{\xi}) \neq 0$, then J belongs to

$$-\Gamma_{\mathbb{Z},\mu}(\mathbf{E}_S) = \Big\{J = (j_1,\ldots,j_s) \in \mathbb{Z}^s : 0 < j_p \leq \tau_S(k_p,k_r)j_r + \mu\Big\}.$$

PROOF. Suppose $\Theta_J(\boldsymbol{\xi})\Psi_S(\boldsymbol{\xi})m(\boldsymbol{\xi}) \neq 0$. Then $\Theta_J(\boldsymbol{\xi}) \neq 0$, and it follows that $2^{j_r-1} \leq \widehat{n}_{S,r}(\boldsymbol{\xi}_{I_r}) \leq 2^{j_r+2}$. Also $\Psi_S(\boldsymbol{\xi}) \neq 0$, and it follows that $\boldsymbol{\xi} \in \widehat{E}_S^A$ and so

$$1 \leq \widehat{n}_{S,p}(\boldsymbol{\xi}_{I_p})^{1/\tau_S(k_p,k_r)} < A\,\widehat{n}_{S,r}(\boldsymbol{\xi}_{I_r})$$

for all $r,p \in \{1,\ldots,s\}$. It follows that

$$\big(2^{j_p-1}\big)^{1/\tau_S(k_p,k_r)} \leq \widehat{n}_{S,p}(\boldsymbol{\xi}_{I_p})^{1/\tau_S(k_p,k_r)} < A\,\widehat{n}_{S,r}(\boldsymbol{\xi}_{I_r}) \leq A\,2^{j_r+1},$$

and so

$$j_p \leq \tau_S(k_p,k_r)j_r + \big[\tau_S(k_p,k_r)(1+\log_2(A)) + 1\big] \leq \tau_S(k_p,k_r)j_r + \mu$$

for an appropriate constant μ. □

It now follows that

$$\begin{aligned} m_S(\boldsymbol{\xi}) &= \sum_{J\in-\Gamma_{\mathbb{Z},\mu}(\mathbf{E}_S)}\Theta_J(\boldsymbol{\xi})m_S(\boldsymbol{\xi}) \\ &= \sum_{J\in-\Gamma_{\mathbb{Z},\mu}(\mathbf{E}_S)}\Theta_{\mathbf{0}}(2^{-j_1}\,\hat{\cdot}_S\,\boldsymbol{\xi}_{I_1},\ldots,2^{-j_s}\,\hat{\cdot}_S\,\boldsymbol{\xi}_{I_s})\,m_S(\boldsymbol{\xi}_{I_1},\ldots,\boldsymbol{\xi}_{I_s}). \end{aligned}$$

Put

$$m_S^J(\boldsymbol{\xi}) = \Theta_0(\boldsymbol{\xi})m_S(2^{j_1}\,\hat{\cdot}_S\,\boldsymbol{\xi}_{I_1},\ldots,2^{j_s}\,\hat{\cdot}_S\,\boldsymbol{\xi}_{I_s})$$

so that we can write

$$m_S(\boldsymbol{\xi}_{I_1},\ldots,\boldsymbol{\xi}_{I_s}) = \sum_{J\in-\Gamma_{\mathbb{Z},\mu}(\mathbf{E}_S)} m_S^J(2^{-j_1}\hat{\cdot}_S\boldsymbol{\xi}_{I_1},\ldots,2^{-j_s}\hat{\cdot}_S\boldsymbol{\xi}_{I_S}).$$

PROPOSITION 6.12. The set of functions $\Big\{m_S^J : J \in -\Gamma_{\mathbb{Z},\mu}(\mathbf{E}_S)\Big\}$ is a uniformly bounded family in $\mathcal{C}_0^\infty(\mathbb{R}^N)$, each supported where $\frac{1}{4} < \widehat{n}_{S,r}(\boldsymbol{\xi}_{I_r}) < 4$ for $1 \leq r \leq s$.

PROOF. Since m_S is supported on $\widehat{E}_S^A$,

$$\begin{aligned} &\big|\partial^\gamma\big(m_S(2^{j_1}\,\hat{\cdot}_S\,\boldsymbol{\xi}_{I_1},\ldots,2^{j_s}\,\hat{\cdot}_S\,\boldsymbol{\xi}_{I_s})\big)(\boldsymbol{\xi})\big| \\ &\qquad = 2^{\sum_{r=1}^s [\![\gamma_r]\!]/e(k_r,j)}\left|\partial_{\boldsymbol{\xi}}^\gamma(m_S)(2^{j_1}\,\hat{\cdot}_S\,\boldsymbol{\xi}_{I_1},\ldots,2^{j_s}\,\hat{\cdot}_S\,\boldsymbol{\xi}_{I_s})\right| \\ &\qquad \lesssim 2^{\sum_{r=1}^s [\![\gamma_r]\!]/e(k_r,j)}\prod_{r=1}^s \widehat{n}_{S,r}(2^{j_r}\,\hat{\cdot}_S\,\boldsymbol{\xi}_{I_r})^{-[\![\gamma_l]\!]/e(k_r,j)} \\ &\qquad = C_\gamma\prod_{r=1}^s \widehat{n}_{S,r}(\boldsymbol{\xi}_{I_r})^{-[\![\gamma_l]\!]/e(k_r,j)}. \end{aligned}$$

However, on the support of $\theta_{\mathbf{0}}$, the product $\prod_{r=1}^s \widehat{n}_{S,r}(\boldsymbol{\xi}_{I_r})^{-[\![\gamma_l]\!]/e(k_r,j)}$ is uniformly bounded and bounded away from zero (depending on γ). □

We have written $m_S(\boldsymbol{\xi}) = \sum_{J\in -\Gamma_{\mathbb{Z},\mu}(\mathbf{E}_S)} m_S^J(2^{-J} \hat{\cdot}_S \boldsymbol{\xi})$. We want to replace the sum over $-\Gamma_{\mathbb{Z},\mu}(\mathbf{E}_S)$ by a sum over $-\Gamma_{\mathbb{Z}}(\mathbf{E}_S)$.

For each $J \in -\Gamma_{\mathbb{Z},\mu}(\mathbf{E}_S)$, choose $J' = p(J) \in -\Gamma_{\mathbb{Z}}(\mathbf{E}_S)$ at minimal distance from J. It is quite clear that there is a finite upper bound M for the number of J having the same $p(J)$ and that, if $p(J) = J'$, then $|j_r - j_r'| \le c(\mu)$ for each $r = 1, \ldots, s$. It follows that we can rewrite

$$\begin{aligned} m_S(\boldsymbol{\xi}) &= \sum_{J\in -\Gamma_{\mathbb{Z},\mu}(\mathbf{E}_S)} m_S^J(2^{-J} \hat{\cdot}_S \boldsymbol{\xi}) \\ &= \sum_{J'\in -\Gamma_{\mathbb{Z}}(\mathbf{E}_S)} \sum_{J:p(J)=J'} m_S^J(2^{-J} \hat{\cdot}_S \boldsymbol{\xi}) \\ &= \sum_{J'\in -\Gamma_{\mathbb{Z}}(\mathbf{E}_S)} \widetilde{m}_S^{J'}(2^{-J'} \hat{\cdot}_S \boldsymbol{\xi}) , \end{aligned} \tag{6.5}$$

where the functions

$$\widetilde{m}_S^{J'}(\boldsymbol{\xi}) = \sum_{J:p(J)=J'} m_S^J(2^{J'-J} \hat{\cdot}_S \boldsymbol{\xi})$$

form a a uniformly bounded family in $\mathcal{C}_0^\infty(\mathbb{R}^N)$.

Applying this decomposition for each marked partition $S \in \mathcal{S}'(\mathbf{E})$ we have the following decomposition for multipliers.

THEOREM 6.13. *Let $m \in \mathcal{M}_\infty(\mathbf{E})$. There is a uniformly bounded collection of functions $\{m_S^J\} \subset \mathcal{C}_0^\infty(\mathbb{R}^N)$ indexed by*

$$S \in \mathcal{S}'(\mathbf{E}) = \big\{T \in \mathcal{S}(n) : \widehat{E}_T \cap \mathbb{B}(1)^c \neq \emptyset\big\}$$

and $J \in \Gamma_{\mathbb{Z}}(\mathbf{E}_S)$ with the following properties.

(a) *If $S = \big\{(I_1, k_1); \ldots; (I_s, k_s)\big\} \in \mathcal{S}'(\mathbf{E})$ and $J \in \Gamma_{\mathbb{Z}}(\mathbf{E}_S)$, the function m_S^J is supported where $c^{-1} < \widehat{n}_{S,r}(\boldsymbol{\xi}_{I_r}) < c$ for $1 \le r \le s$ and an appropriate constant c.*

(b) *The multiplier m can be written*

$$m(\boldsymbol{\xi}) = m_0(\boldsymbol{\xi}) + \sum_{S\in\mathcal{S}'(\mathbf{E})} \sum_{J\in -\Gamma_{\mathbb{Z}}(\mathbf{E}_S)} m_S^J(2^{-j_1} \hat{\cdot}_S \boldsymbol{\xi}_{i_1}, \ldots, 2^{-j_s} \hat{\cdot}_S \boldsymbol{\xi}_{I_s}) \tag{6.6}$$

where $m_0 \in \mathcal{C}_0^\infty(\mathbb{R}^N)$ is supported in $\mathbb{B}(2)$.

6.5. Dyadic decomposition of a kernel

We can now decompose a distribution $\mathcal{K} \in \mathcal{P}_0(\mathbf{E})$. We know from Theorem 4.2 that the Fourier transform $m = \widehat{\mathcal{K}} \in \mathcal{M}_\infty(\mathbf{E})$, and by Theorem 6.13 we can write m as in equation (6.6). Write

$$[\psi]^J(\boldsymbol{\xi}_{I_1}, \ldots, \boldsymbol{\xi}_{I_s}) = \psi\big(2^{-j_1} \hat{\cdot}_S \boldsymbol{\xi}_{I_1}, \ldots, 2^{-j_s} \hat{\cdot}_S \boldsymbol{\xi}_{I_s}\big)$$

so that

$$\widehat{\mathcal{K}} = m_0 + \sum_{S\in\mathcal{S}'(\mathbf{E})} \sum_{J\in -\Gamma_{\mathbb{Z}}(\mathbf{E}_S)} [m_S^J]^J.$$

Then if $^\vee$ denotes the inverse Fourier transform, we have

$$\mathcal{K} = (m_0)^\vee + \sum_{S\in\mathcal{S}'(\mathbf{E})} \sum_{J\in -\Gamma_{\mathbb{Z}}(\mathbf{E}_S)} ([m_S^J]^J)^\vee.$$

Computing the inverse Fourier transform of $[m_S^J]^J$ we have

$$([m_S^J]^J)^\vee(\mathbf{x}_{I_1},\ldots,\mathbf{x}_{I_S}) = 2^{\sum_{r=1}^s j_r \widehat{Q}_{I_r}}\,[m_S^J]^\vee(2^{j_1}\hat{\cdot}_S\,\mathbf{x}_{I_1},\ldots,2^{j_s}\hat{\cdot}_S\,\mathbf{x}_{I_s}).$$

Since the family $\{m_S^J\}$ is uniformly bounded in $\mathcal{C}_0^\infty(\mathbb{R}^N)\subset\mathcal{S}(\mathbb{R}^N)$, it follows that the family $\{([m_S^J]^J)^\vee\}$ is also a uniformly bounded family in $\mathcal{S}(\mathbb{R}^N)$. Since each m_S^J is supported where $c^{-1}<\widehat{n}_{S,r}(\boldsymbol{\xi}_{I_r})<c$ for all r, and since $\widehat{n}_{S,r}(\boldsymbol{\xi}_{I_r})\approx n_{k_r}(\boldsymbol{\xi}_{k_r})$, it follows that

$$\int_{\mathbb{R}^{C_{k_r}}}(m_S^J)^\vee(\mathbf{x}_1,\ldots,\mathbf{x}_{k_r},\ldots,\mathbf{x}_n)\,d\mathbf{x}_{k_r}=0 \tag{6.7}$$

for $1\le r\le s$; *i.e.* the function $(m_S^J)^\vee$ has cancellation in each of the variables $\{\mathbf{x}_{k_1},\ldots,\mathbf{x}_{k_s}\}$. Let us write

$$\psi_0=(m_0)^\vee,\qquad \psi_S^J=([m_S^J]^J)^\vee=[(m_S^{-J})^\vee]_{-J}. \tag{6.8}$$

THEOREM 6.14. *Let $\mathcal{K}\in\mathcal{P}_0(\mathbf{E})$. There is a function $\psi_0\in\mathcal{S}(\mathbb{R}^N)$ and for each $S\in\mathcal{S}'(\mathbf{E})$ there is a uniformly bounded collection of functions $\{\psi_S^J\}_{J\in\Gamma_{\mathbb{Z}}(\mathbf{E}_S)}\subset\mathcal{S}(\mathbb{R}^N)$ so that:*

1. *Let $S=\{(I_1,k_1);\ldots;(I_s,k_s)\}\in\mathcal{S}'(\mathbf{E})$.*
   (a) *Each function ψ_S^J has cancellation in each of the variables $\{\mathbf{x}_{k_1},\ldots,\mathbf{x}_{k_s}\}$, and so has strong cancellation relative to the decomposition $\mathbb{R}^N=\mathbb{R}^{I_1}\oplus\cdots\oplus\mathbb{R}^{I_s}$.*
   (b) *The series*
   $$\mathcal{K}_S(\mathbf{x})=\sum_{J\in\Gamma_{\mathbb{Z}}(\mathbf{E}_S)}[\psi_S^J]_J(\mathbf{x})=\sum_{J\in\Gamma_{\mathbb{Z}}(\mathbf{E}_S)}2^{-\sum_{r=1}^s i_r\widehat{Q}_{I_r}}\psi_S^J(2^{-J}\hat{\cdot}_S\mathbf{x})$$
   converges to an element of $\mathcal{P}_0(\mathbf{E}_S)\subset\mathcal{P}_0(\mathbf{E})$ in the sense of distributions .
2. *The distribution $\mathcal{K}$ decomposes as $\mathcal{K}(\mathbf{x})=\psi_0(\mathbf{x})+\sum_{S\in\mathcal{S}'(\mathbf{E})}\mathcal{K}_S(\mathbf{x})$.*

PROOF. Let $m=\widehat{\mathcal{K}}\in\mathcal{M}_\infty(\mathbf{E})$. Using Theorem 6.13, write

$$m=m_0+\sum_{S\in\mathcal{S}'(\mathbf{E})}\sum_{J\in-\Gamma_{\mathbb{Z}}(\mathbf{E}_S)}[m_S^J]^J.$$

Then for each $S\in\mathcal{S}'(\mathbf{E})$ and each $J\in\Gamma_{\mathbb{Z}}(\mathbf{E}_S)$, let $\psi_S^J(\mathbf{x})=(m_S^{-J})^\vee(\mathbf{x})$. We have already observed in equation (6.7) that ψ_S^J has cancellation in each variable $\mathbf{x}_{k_r}$, $1\le r\le s$, which gives assertion (1a).

Next

$$\mathcal{K}_S-\sum_{J\in\Gamma_{\mathbb{Z}}(\mathbf{E}_S)}[\psi_S^J]_J=\sum_{J\in-\Gamma_{\mathbb{Z}}(\mathbf{E}_S)}[\psi_S^{-J}]_{-J}=\sum_{J\in-\Gamma_{\mathbb{Z}}(\mathbf{E}_S)}([m_S^J]^J)^\vee=m_S^\vee.$$

According to Proposition 6.10, $m_S\in\mathcal{M}_\infty(\mathbf{E}_S)$, and using Theorem 4.1 it follows that $\mathcal{K}_S\in\mathcal{P}_0(\mathbf{E}_S)$. But according to Lemma 6.4, $\mathcal{M}_\infty(\mathbf{E}_S)\subset\mathcal{M}_\infty(\mathbf{E})$, and every distribution in $\mathcal{P}_0(\mathbf{E}_S)$ belongs to $\mathcal{P}_0(\mathbf{E})$. This establishes assertion (1b).

Finally assertion (2) is a consequence of the identity $m=m_0+\sum_{S\in\mathcal{S}'(\mathbf{E})}m_S$, and this completes the proof. □

Theorem 6.14 shows that each kernel $\mathcal{K}$ can be written as sums of dilates of uniformly bounded families of Schwartz functions. As in [**NRSW12**], we can improve this result and show that we can replace Schwartz functions by uniformly bounded families of functions compactly supported in the unit ball. We use the following result, which is Lemma 6.5 in [**NRSW12**].

LEMMA 6.15. *Let $\psi \in \mathcal{S}(\mathbb{R}^N)$. Then there are functions $\{\varphi^k\} \subset C_0^\infty(\mathbb{R}^N)$, $k \in \mathbb{N}$ such that*

$$\psi(\mathbf{x}) = \sum_{k=0}^{\infty} 2^{-kQ} \varphi^k(2^{-k} \cdot \mathbf{x})$$

where Q is the homogeneous dimension of $\mathbb{R}^N$ relative to the given dilations. The functions $\{\varphi^k\}$ have the following properties.

(a) *Each φ^k is supported in $\mathbb{B}(1)$.*

(b) *For any $\delta > 0$ and any $\gamma \in \mathbb{N}^N$ there exists a positive integer M so that*

$$\sup_{\mathbf{x}\in\mathbb{R}^N} |\partial^\gamma \varphi^k(\mathbf{x})| \leq 2^{-k\delta} \sup_{\mathbf{x}\in\mathbb{R}^N} \sum_{|\boldsymbol{\alpha}|\leq M} |\partial^{\boldsymbol{\alpha}} \psi(\mathbf{x})|.$$

(c) *If ψ has strong cancellation, then each φ^k also has strong cancellation.*

Now let $\mathcal{K} \in \mathcal{P}_0(\mathbf{E})$ and let $\mathcal{K} = \psi_0 + \sum_{S\in\mathcal{S}'(\mathbf{E})} \mathcal{K}_S$ be the decomposition given in Theorem 6.14 where $\mathcal{K}_S(\mathbf{x}) = \sum_{J\in\Gamma_{\mathbb{Z}}(\mathbf{E}_S)} [\psi_S^J]_J(\mathbf{x})$, and each ψ_S^J has cancellation in the variables $\{\mathbf{x}_{k_1}, \dots, \mathbf{x}_{k_s}\}$. We apply Lemma 6.15 to each function ψ_S^J:

$$\psi_S^J(\mathbf{x}) = \sum_{k=0}^{\infty} 2^{-kQ_S} \varphi_S^{J,k}(2^{-k} \cdot \mathbf{x})$$

where each $\{\varphi_S^{J,k}\} \subset C_0^\infty(\mathbb{R}^N)$ is a uniformly bounded family of functions supported in $\mathbb{B}(1)$, each having cancellation in the variables $\{\mathbf{x}_{k_1}, \dots, \mathbf{x}_{k_s}\}$. Then

$$\mathcal{K}_S(\mathbf{x}) = \sum_{J\in\Gamma_{\mathbb{Z}}(\mathbf{E}_S)} \left[\sum_{k=0}^{\infty} [\varphi_S^{J,k}]_k\right]_J (\mathbf{x}).$$

The argument on pages 661-663 of [**NRSW12**] shows that this last sum can be rewritten to yield

$$\mathcal{K}_S(\mathbf{x}) = \sum_{J\in\Gamma_{\mathbb{Z}}(\mathbf{E}_S)} [\widetilde{\varphi}_S^J]_J(\mathbf{x})$$

where $\{\widetilde{\varphi}_S^J\}$ is a uniformly bounded family in $C_0^\infty(\mathbb{R}^N)$, all supported in $\mathbb{B}(1)$, and all having cancellation in the variables $\{\mathbf{x}_{k_1}, \dots, \mathbf{x}_{k_s}\}$. Thus we have

COROLLARY 6.16. Let $\mathcal{K} \in \mathcal{P}_0(\mathbf{E})$. For each $S \in \mathcal{S}'(\mathbf{E})$ there is a uniformly bounded collection of functions $\{\varphi_S^J\}_{J\in\Gamma_{\mathbb{Z}}(\mathbf{E}_S)} \subset C_0^\infty(\mathbb{R}^N)$ and a function $\psi_0 \in \mathcal{S}(\mathbb{R}^N)$ so that

(a) Each function φ_S^J has cancellation in the variables $\mathbf{x}_{k_1}, \dots, \mathbf{x}_{k_s}$.

(b) Let

$$[\varphi_S^J]^J(\mathbf{x}) = 2^{-} \varphi_S^J(2^{-j_1} \hat{\cdot}_S \mathbf{x}_{I_1}, \dots, 2^{-j_s} \hat{\cdot}_S \mathbf{x}_{I_s}),$$

where

$$2^{-j_r} \hat{\cdot}_S \mathbf{x}_{I_r} = \left\{2^{-j_r/e(k_r,j)} \cdot \mathbf{x}_j\right\}_{j\in I_r}.$$

Then the series $\mathcal{K}_S = \sum_{J \in \Gamma_{\mathbb{Z}}(\mathbf{E}_S)} \left[\varphi_S^J\right]_J$ converges in the sense of distributions to an element $\mathcal{K}_S \in \mathcal{P}_0(\mathbf{E}_S) \subset \mathcal{P}_0(\mathbf{E})$.

(c) $\mathcal{K}(\mathbf{x}) = \psi_0(\mathbf{x}) + \sum_{S \in \mathcal{S}'(\mathbf{E})} \mathcal{K}_S(\mathbf{x})$.

CHAPTER 7

The rank of E and integrability at infinity

So far we have analyzed kernels in the restricted class $\mathcal{P}_0(\mathbf{E})$, which are the sum of a kernel in $\mathcal{P}(\mathbf{E})$ with compact support and a Schwartz function. In this Section we discuss the nature of a distribution $\mathcal{K} \in \mathcal{P}(\mathbf{E})$ and show that it depends in the first place on the rank of the matrix $\mathbf{E} = \{e(j,k)\}$. We will see that if the rank equals 1 then the distribution $\mathcal{K}$ is a standard (non-isotropic) Calderón-Zygmund kernel. However, if the rank is greater than 1, we will see that the kernel K is integrable at infinity, but has worse behavior near the origin than a Calderón-Zygmund kernel.

We will also show that we have integrability at infinity in a more general situation. Suppose that the Fourier transform m of a distribution $\mathcal{K}$ satisfies the differential inequalities in Definition 2.4 for pure derivatives; *i.e.* for every $1 \le j \le n$ and for every $\boldsymbol{\gamma}_j \in \mathbb{N}^{C_j}$ there is a constant $C_{\boldsymbol{\gamma}_j} > 0$ so that $|\partial_{\boldsymbol{\xi}_j}^{\boldsymbol{\gamma}_j} m(\boldsymbol{\xi})| \le C_{\boldsymbol{\gamma}_j} N_j(\boldsymbol{\xi})^{-[\![\boldsymbol{\gamma}_j]\!]}$. Then if the rank of $\mathbf{E}$ is greater than 1, the function K is integrable at infinity. This result will be used in Section 11 below when we consider distributions which are simultaneously flag kernels for two different flags.

7.1. The rank 1 case: Calderón-Zygmund kernels

We begin by recalling the definition of a Calderón-Zygmund kernel[1]. Start with the usual decomposition

$$\mathbb{R}^N = \mathbb{R}^{C_1} \oplus \cdots \oplus \mathbb{R}^{C_n}, \tag{7.1}$$

and use the standard notation of Section 2. For $\mathbf{a} = (a_1, \ldots, a_n)$ an n-tuple of positive numbers, define a one-parameter family of dilations by setting

$$\lambda \cdot_{\mathbf{a}} \mathbf{x} = (\lambda^{a_1} \cdot \mathbf{x}_1, \ldots, \lambda^{a_n} \cdot \mathbf{x}_n). \tag{7.2}$$

With this homogeneity, $\mathbb{R}^N$ has homogeneous dimension $Q_{\mathbf{a}} = \sum_{j=1}^n a_j Q_j$, and

$$N_{\mathbf{a}}(\mathbf{x}) = n_1(\mathbf{x}_1)^{1/a_1} + \cdots + n_n(\mathbf{x}_n)^{1/a_n}$$

is a homogeneous norm. We can then define the class of Calderón-Zygmund kernels with this homogeneity.

DEFINITION 7.1. $\mathcal{CZ}_{\mathbf{a}}$ is the space of tempered distributions $\mathcal{K}$ on $\mathbb{R}^N$ such that

(a) away from the origin, $\mathcal{K}$ is given by integration against a smooth function K satisfying the differential inequalities

$$\big|\partial^{\boldsymbol{\gamma}} K(\mathbf{x})\big| \le C_{\boldsymbol{\gamma}} N_{\mathbf{a}}(\mathbf{x})^{-Q_{\mathbf{a}} - [\![\boldsymbol{\gamma}]\!]_{\mathbf{a}}},$$

for every multi-index $\boldsymbol{\gamma}$, with $[\![\boldsymbol{\gamma}]\!]_{\mathbf{a}} = \sum_{j=1}^n a_j [\![\boldsymbol{\gamma}_j]\!]$;

[1] For reasons of compatibility with the framework of this paper, we restrict ourselves to kernels and multipliers which are C^∞ away from the origin and to dilations which are compatible with the decomposition 7.1 and with the dilations in (2.5).

(b) there is a constant $C > 0$ so that for any normalized bump function $\psi \in C_0^\infty(\mathbb{R}^N)$ with support in the unit ball and for any $R > 0$, $\left|\langle \mathcal{K}, \psi_R \rangle\right| \le C$, where $\psi_R(\mathbf{x}) = \psi(R \cdot_{\mathbf{a}} \mathbf{x})$.

It is well known, cf. [**NRS01**], that, for distributions $\mathcal{K} \in \mathcal{CZ}_{\mathbf{a}}$, the corresponding multipliers $m = \widehat{\mathcal{K}}$ are characterized by the differential inequalities

$$\left|\partial^\gamma m(\boldsymbol{\xi})\right| \le C_\gamma N_{\mathbf{a}}(\boldsymbol{\xi})^{-[\![\gamma]\!]_{\mathbf{a}}}.$$

Put

$$N_j(\boldsymbol{\xi}) = n_1(\boldsymbol{\xi})^{\frac{a_j}{a_1}} + \cdots + n_n(\boldsymbol{\xi})^{\frac{a_j}{a_n}} \approx N_{\mathbf{a}}(\boldsymbol{\xi})^{a_j}.$$

Then these differential inequalities can be written as

$$\left|\partial^\gamma m(\boldsymbol{\xi})\right| \le C_\gamma \prod_{j=1}^n N_j(\boldsymbol{\xi})^{-[\![\gamma_j]\!]},$$

and so the multipliers are the elements of the class $\mathcal{M}(\mathbf{E}_{\mathbf{a}})$ where

$$\mathbf{E}_{\mathbf{a}} = \begin{bmatrix} 1 & \frac{a_1}{a_2} & \frac{a_1}{a_3} & \cdots & \frac{a_1}{a_n} \\ \frac{a_2}{a_1} & 1 & \frac{a_2}{a_3} & \cdots & \frac{a_2}{a_n} \\ \vdots & \vdots & \vdots & \ddots & \vdots \\ \frac{a_n}{a_1} & \frac{a_n}{a_2} & \frac{a_n}{a_3} & \cdots & 1 \end{bmatrix}. \tag{7.3}$$

PROPOSITION 7.2. Let $\mathbf{E} = \{e(j,k)\}$ satisfy the basic hypotheses (2.10). The following are equivalent:

(1) $\operatorname{rank}(\mathbf{E}) = 1$;

(2) $e(j,k) = e(j,l)e(l,k)$ for all $1 \le j,k,l \le n$;

(3) there is a dilation structure on $\mathbb{R}^N$, compatible with the decomposition (7.1), so that if $m \in \mathcal{M}(\mathbf{E})$, then m is a Mihlin-Hörmander multiplier relative to that structure;

(4) there is a dilation structure on $\mathbb{R}^N$, compatible with the decomposition (7.1), so that if $\mathcal{K} \in \mathcal{P}(\mathbf{E})$, then $\mathcal{K}$ is a Calderón-Zygmund kernel relative to that structure.

PROOF. The rank of $\mathbf{E}$ equals 1 if and only if all the rows are proportional. Due to the 1's along the diagonal this means that, for every $j \ne k$ the j-th row equals the k-th row multiplied by $e(j,k)$. Thus (1) and (2) are equivalent.

If (2) holds, we have in particular that $e(j,k)e(k,j) = 1$ for all j,k. Hence $\widehat{N}_j = N_j$ for all j. Moreover, in analogy with (5.6),

$$N_j(\mathbf{x}) \approx N_1(\mathbf{x})^{e(j,1)}.$$

Thus the differential inequalities in Definition 2.4 can be written

$$\begin{aligned} \left|\partial^\gamma m(\boldsymbol{\xi})\right| &\le C_\gamma \prod_{j=1}^n \widehat{N}_j(\boldsymbol{\xi})^{-[\![\gamma_j]\!]} \\ &\approx \prod_{j=1}^n N_1(\boldsymbol{\xi})^{-e(j,1)[\![\gamma_j]\!]} = N_1(\boldsymbol{\xi})^{-\sum_{j=1}^n e(j,1)[\![\gamma_j]\!]} \\ &= N_1(\boldsymbol{\xi})^{-[\![\gamma]\!]}, \end{aligned}$$

where $[\![\gamma]\!] = \sum_{j=1}^{n} e(j,1)[\![\gamma_j]\!] = \sum_{j=1}^{n} e(1,j)^{-1}[\![\gamma_j]\!]$ is the length of γ in the dilation structure that makes N_1 a homogeneous norm. Thus the multiplier $m(\boldsymbol{\xi})$ satisfies the Mihlin-Hörmander differential inequalities relative to the family of dilations, and so (3) holds.

Conversely, assume that (3) holds. Then there are exponents $\beta_1 = 1, \beta_2, \ldots, \beta_n$ such that, setting

$$N(\boldsymbol{\xi}) = \sum_{j=1}^{n} n_j(\boldsymbol{\xi})^{\beta_j},$$

every $m \in \mathcal{M}(\mathbf{E})$ satisfies the inequalities

$$\left|\partial^{\gamma} m(\boldsymbol{\xi})\right| \lesssim N(\boldsymbol{\xi})^{-\sum_{j=1}^{n} \beta_j^{-1}[\![\gamma_j]\!]}.$$

This is equivalent to saying that $m \in \mathcal{M}(\mathbf{E}')$ with

$$e'(j,k) = \frac{\beta_k}{\beta_j}.$$

Hence in our hypothesis we have the inclusion $\mathcal{M}(\mathbf{E}) \subseteq \mathcal{M}(\mathbf{E}')$, which implies that $\mathcal{M}_\infty(\mathbf{E}) \subseteq \mathcal{M}_\infty(\mathbf{E}')$. By Proposition 5.2, this implies that $\Gamma(\mathbf{E}) \subseteq \Gamma(\mathbf{E}')$. By Lemma 14.9, the dimension of $\Gamma(\mathbf{E}')$ equals the reduced rank of $\mathbf{E}'$, which is 1. Hence also $\Gamma(\mathbf{E})$ has dimension 1, so it must coincide with $\Gamma(\mathbf{E}')$. This proves that (3) implies (1).

Finally, the equivalence of (3) and (4) is proved in [**NRS01**]. □

7.2. Higher rank and integrability at infinity

In this section let $\mathbf{E} = \{e(j,k)\}$ be an $n \times n$ matrix satisfying the basic hypothesis (2.10), and as usual let

$$N_j(\mathbf{x}) = \sum_{k=1}^{n} n_k(\mathbf{x}_k)^{e(j,k)} \qquad \text{and} \quad \widehat{N}_j(\boldsymbol{\xi}) = \sum_{k=1}^{n} n_k(\boldsymbol{\xi}_k)^{1/e(k,j)}.$$

Let Q_j be the homogeneous dimension of $\mathbb{R}^{C_j}$. We begin by studying the integrability at infinity of a measurable function K on $\mathbb{R}^N$ which satisfies the inequality

$$|K(\mathbf{x})| \leq C \prod_{j=1}^{n} N_j(\mathbf{x})^{-Q_j}. \tag{7.4}$$

Note that nothing is said about the size of the derivatives of K.

LEMMA 7.3. *Suppose that the rank of* $\mathbf{E}$ *is greater than or equal to* 2. *If* K *satisfies the inequality in equation (7.4), then* K *is integrable over the complement of the unit ball* $\mathbb{B}(1)$.

PROOF. Since the rank of $\mathbf{E}$ is strictly larger than 1, it follows from Proposition 7.2 that there are integers $j,k,l \in \{1,\ldots,n\}$ so that $e(j,k) < e(j,l)e(l,k)$. Moreover, for every $k \in \{1,\ldots,n\}$, we have

$$\begin{aligned} |K(\mathbf{x})| &\leq C \prod_{j=1}^{n} N_j(\mathbf{x})^{-Q_j} \leq C \prod_{j=1}^{n} n_k(\mathbf{x}_k)^{-e(j,k)Q_j} \\ &= C n_k(\mathbf{x}_k)^{-\sum_{j=1}^{n} e(j,k)Q_j} = C n_k(\mathbf{x}_k)^{-\widetilde{Q}_k} \end{aligned}$$

where $\widetilde{Q}_k = \sum_{j=1}^{n} e(j,k)Q_j$. Decompose the complement of $\mathbb{B}(1)$ as the union of sets E_ω where ω ranges over the collection of non-empty subsets of $\{1,\ldots,n\}$, and

$$E_\omega = \{\mathbf{x}\in\mathbb{R}^N : n_k(\mathbf{x}_k)\geq 1 \text{ for } k\in\omega,\ n_k(\mathbf{x}_k)<1 \text{ for } k\notin\omega\}.$$

Fix ω, and let $\{\theta_l\}_{l\in\omega}$ be strictly positive numbers such that $\sum_{l\in\omega}\theta_l = 1$. Then it follows from (7.2) that $|K(\mathbf{x})| \lesssim \prod_{l\in\omega}|\mathbf{x}_l|_l^{-\theta_l\widetilde{Q}_l}$, and if we can choose the constants $\{\theta_l\}$ such that $\theta_l\widetilde{Q}_l > Q_l$ for every l, then

$$\int_{E_\omega}|K(\mathbf{x})|\,d\mathbf{x} \lesssim \prod_{l\in\omega}\int_{\substack{\mathbf{x}_l\in\mathbb{R}^{C_l}\\ |\mathbf{x}_l|\geq 1}}|\mathbf{x}_l|^{-\theta_l\widetilde{Q}_l}\,d\mathbf{x}_l < +\infty.$$

We can choose $\{\theta_l\}$ with the required property if $\sum_{l\in\omega}Q_l\widetilde{Q}_l^{-1} < 1$, and we can do this for every subset ω if $\sum_{l=1}^{n}Q_l\widetilde{Q}_l^{-1} < 1$. But for any fixed $k\in\{1,\ldots,n\}$, we have $e(j,l)\geq e(j,k)/e(l,k)$ and so

$$\sum_{l=1}^{n}Q_l\widetilde{Q}_l^{-1} = \sum_{l=1}^{n}Q_l\Big[\sum_{j=1}^{n}e(j,l)Q_j\Big]^{-1} \leq \sum_{l=1}^{n}e(l,k)\,Q_l\Big[\sum_{j=1}^{n}e(j,k)Q_j\Big]^{-1} = 1. \tag{7.5}$$

Moreover, we can make the inequality in (7.5) strict if there exists a single triple $j,k,l\in\{1,\ldots,n\}$ such that $e(j,k) < e(j,l)e(l,k)$. This completes the proof. □

COROLLARY 7.4. Suppose that $\operatorname{rank}(\mathbf{E})\geq 2$. If $\mathcal{K}\in\mathcal{P}(\mathbf{E})$ we can write $\mathcal{K} = \mathcal{K}_0 + \mathcal{K}_\infty$ where $\mathcal{K}_0,\mathcal{K}_\infty\in\mathcal{P}(\mathbf{E})$, $\mathcal{K}_0$ is a distribution with compact support in the unit ball, and $\mathcal{K}_\infty$ is a distribution given by integration against a function $K_\infty\in\mathcal{C}^\infty(\mathbb{R}^N)\cap L^1(\mathbb{R}^N)$ which satisfies the differential inequalities of Definition 2.2.

We also will need the following variant of Lemma 7.3.

LEMMA 7.5. *Suppose that $\mathcal{K}$ is a tempered distribution on $\mathbb{R}^n$ and suppose that its Fourier transform $m = \widehat{\mathcal{K}}$ is a smooth function away from zero which satisfies the following differential inequalities for pure derivatives: for $1\leq j\leq n$ and for every $\boldsymbol{\gamma}_j\in\mathbb{N}^{C_j}$ there exists a constant $C_{\boldsymbol{\gamma}_j}>0$ so that*

$$\big|\partial_{\boldsymbol{\xi}_j}^{\boldsymbol{\gamma}_j}m(\boldsymbol{\xi})\big| \leq C_{\boldsymbol{\gamma}_j}\widehat{N}_j(\boldsymbol{\xi})^{-[\![\boldsymbol{\gamma}_j]\!]}.$$

Then $\mathcal{K}$ is given by integration against a smooth function K away from the origin, and if the rank of the matrix $\mathbf{E}$ is greater than 1, then K is integrable at infinity.[2]

PROOF. If we can show that $\mathcal{K}$ is given by integration agains a function K with $|K(\mathbf{x})|\lesssim n_k(\mathbf{x}_k)^{-\sum_{j=1}^{n}e(j,k)Q_j}$ for $1\leq k\leq n$, then the argument proceeds as in the proof of Lemma 7.3. Thus fix k, and define $\lambda\cdot_k\mathbf{x} = (\lambda^{e(1,k)}\cdot\mathbf{x}_1,\ldots,\lambda^{e(n,k)}\cdot\mathbf{x}_n)$. This is a family of dilations on all of $\mathbb{R}^N$ and $\widehat{N}_j(\mathbf{t}) = \sum_{k=1}^{n}n_k(\mathbf{t}_k)^{1/e(k,j)}$ is a homogeneous norm. The homogeneous dimension is $\sum_{j=1}^{n}e(j,k)Q_j$. Choose $\chi\in\mathcal{C}_0^\infty(\mathbb{R}^N)$ with $\chi(\boldsymbol{\xi})\equiv 1$ if $\widehat{N}_j(\boldsymbol{\xi})\leq 1$. Let $m_R(\boldsymbol{\xi}) = \chi(R\cdot_j\boldsymbol{\xi})m(\boldsymbol{\xi})$. Then for all $\boldsymbol{\gamma}_j\in\mathbb{N}^{C_j}$, the product and chain rules show that we have

$$\big|\partial_{\boldsymbol{\xi}_j}^{\boldsymbol{\gamma}_j}m_R(\boldsymbol{\xi})\big| \leq C_{\boldsymbol{\gamma}_j}\widehat{N}_j(\boldsymbol{\xi})^{-[\![\boldsymbol{\gamma}_j]\!]}.$$

[2]Note that we only make an hypothesis about "pure" derivatives of m. We are not assuming that $m\in\mathcal{M}(\mathbf{E})$.

Now $m_R \to m$ in the sense of distributions as $R \to 0$, and hence $\mathcal{K}$ is the limit, in the sense of distributions, of the inverse Fourier transform of m_R. But it now follows from Proposition 15.4 in Appendix B that $|K(\mathbf{x})| \lesssim n_k(\mathbf{x}_k)^{-\sum_{j=1}^n e(j,k)Q_j}$, and this completes the proof. □

7.3. Higher rank and weak-type estimates near zero

If K is the smooth function on $\mathbb{R}^N \setminus \{0\}$ corresponding to a Calderón-Zygmund kernel $\mathcal{K}$, then K satisfies the estimate $\big|\{\mathbf{x} \in \mathbb{R}^N : |K(x)| > \lambda\}\big| \lesssim \lambda^{-1}$. Moreover, there are Calderón-Zygmund kernels $\mathcal{K}$ such that $\big|\{\mathbf{x} \in \mathbb{R}^N : |K(x)| > \lambda\}\big| \gtrsim \lambda^{-1}$. In this section we show that the corresponding estimates for kernels $\mathcal{K} \in \mathcal{P}(\mathbf{E})$ associated with the decomposition $\mathbb{R}^N = \mathbb{R}^{C_1} \oplus \cdots \oplus \mathbb{R}^{C_n}$ depend on the reduced rank of $\mathbf{E}$. Recall from the discussion in Section 5.3 and Proposition 5.5 that if the reduced rank of $\mathbf{E}$ is $m < n$ then there is an $m \times m$ matrix $\mathbf{E}^\flat$ so that $\mathcal{P}_0(\mathbf{E})$ coincides with the space of distributions $\mathcal{P}_0(\mathbf{E}^\flat)$ associated with a coarser decomposition decomposition $\mathbb{R}^N = \mathbb{R}^{A_1} \oplus \cdots \oplus \mathbb{R}^{A_m}$

LEMMA 7.6. *Suppose that the reduced rank of* $\mathbf{E}$ *is* m.

(a) *If* $\mathcal{K} \in \mathcal{P}(\mathbf{E})$, *then for* $\lambda \geq 1$, $\left|\left\{\mathbf{x} \in \mathbb{R}^N : K(\mathbf{x}) > \lambda\right\}\right| \lesssim \lambda^{-1} \log(\lambda)^{m-1}$.

(b) *There exists* $\mathcal{K} \in \mathcal{P}(\mathbf{E})$ *so that* $\left|\left\{\mathbf{x} \in \mathbb{R}^N : K(\mathbf{x}) > \lambda\right\}\right| \gtrsim \lambda^{-1} \log(\lambda)^{m-1}$ *for* $\lambda \geq 1$.

If $m = 1$, we are dealing with a Calderón-Zygmund kernel and these facts are well-known, so we may assume $m > 1$.

We prove parts (a) and (b) of Lemma7.6 separately, since the proof of part (a) requires an intermediate calculation.

PROOF OF LEMMA 7.6, PART (a). As indicated above, we can assume that $m = n$ by replacing $\mathbf{E}$ by its reduced matrix and applying Proposition 5.5. Suppose that $\lambda \geq 1$. For $1 \leq k \leq n$ we have $\prod_{j=1}^n N_j(\mathbf{x})^{Q_j} \geq n_k(\mathbf{x}_k)^{\sum_{l=1}^n e(l,k)Q_l}$, and also $\prod_{j=1}^n N_j(\mathbf{x})^{Q_j} \geq \prod_{l=1}^n n_l(\mathbf{x}_l)^{Q_l}$. Since $|K(\mathbf{x})| \leq \prod_{j=1}^n N_j(\mathbf{x})^{-Q_j}$, it follows that

$$\begin{aligned}&\left\{\mathbf{x} \in \mathbb{R}^N : |K(\mathbf{x})| > \lambda\right\}\\ &\qquad \subset \left\{\mathbf{x} \in \mathbb{R}^N : \begin{cases}\prod_{l=1}^n n_l(\mathbf{x}_l)^{Q_l} < \lambda^{-1} \text{ and} & \\ n_l(\mathbf{x}_l)^{Q_l} < \lambda^{-Q_l/\sum_{k=1}^n e(k,l)Q_k} & 1 \leq l \leq n\end{cases}\right\}.\end{aligned}$$

Note that, as in equation (7.5), $\sum_{l=1}^n Q_l \left[\sum_{k=1}^n e(k,l)Q_k\right]^{-1} < 1$ since the rank of $\mathbf{E}$ is strictly greater than 1. Part (a) of the Lemma then follows from the following calculation. □

PROPOSITION 7.7. Let $A_1, \ldots, A_n, \delta \in (0,\infty)$ and put $A = \prod_{j=1}^n A_j$. Put

$$\begin{aligned}&E(A_1, \ldots, A_n,\delta)\\ &\qquad = \left\{\mathbf{x} \in \mathbb{R}^N : \prod_{j=1}^n n_j(\mathbf{x}_j)^{Q_j} < \delta \text{ and } n_j(\mathbf{x}_j)^{Q_j} < A_j \text{ for } 1 \leq j \leq n\right\}.\end{aligned}$$

There is a constant c_n depending only on n so that

$$\big|E(A_1,\ldots,A_n,\delta)\big| \lesssim c_n \begin{cases} A & \text{if } A<\delta, \\ \delta\left[1+\left(\log\left(\frac{A}{\delta}\right)\right)^{n-1}\right] & \text{if } A\geq\delta \end{cases}$$

PROOF. We argue by induction on n, with the case $n=1$ following from Proposition 15.2 in Appendix B. Thus suppose the result is true for $n-1$. We divide the region $E(A_1,\ldots,A_n,\delta)$ into two parts; the part where $n_n(\mathbf{x}_n)^{Q_n} < \delta[A_1\cdots A_{n-1}]^{-1}$ is a product of n balls, and its volume is δ. For the region where $n_n(\mathbf{x}_n)^{Q_n} \geq \delta[A_1\cdots A_{n-1}]^{-1}$ we can use the induction hypothesis in the variables $\{\mathbf{x}_1,\ldots,\mathbf{x}_{n-1}\}$. □

PROOF OF LEMMA 7.6, PART (b). As before we assume that $\mathbf{E}$ is reduced, i.e., $m=n$. Recall that $\lambda\cdot\mathbf{x} = (\lambda^{d_1}x_1,\ldots,\lambda^{d_N}x_N)$. Choose a closed interval $J_k\subset(1,2^{d_k})\subset\mathbb{R}$, and let $\theta_k\in\mathcal{C}_0^\infty(\mathbb{R})$ be an odd function such that $\theta_k(t)\geq 0$ if $t\geq 0$, $\theta_k(t)\equiv 0$ if $t\notin(1,2^{d_k})$, and $\theta_k(t)\equiv 1$ if $t\in J_k$. Then $\int_{\mathbb{R}}\theta_k(t)\,dt=0$ and if we put $\theta_{k,m}(t)=\theta_k(2^{md_k}t)$, the functions θ_{k,m_1} and θ_{k,m_2} have disjoint supports if $m_1\neq m_2$. Put $\Theta(\mathbf{t})=\prod_{k=1}^N\theta_k(t_k)$.

The function Θ has strong cancellation in the sense of Definition 5.8. It follows from Theorem 5.13 that $\mathcal{K}=\sum_{I\in\Gamma_{\mathbb{Z}}(\mathbf{E})}[\Theta]_I$ converges in the sense of distributions to an element $\mathcal{K}\in\mathcal{P}_0(\mathbf{E})$. Note that the supports of the functions $[\Theta]_I$ are disjoint. Let Σ denote the set where $\Theta=1$, so that if $2^{-I}\cdot\mathbf{x}\in\Sigma$ and $2^{-\sum_{j=1}^n i_jQ_j}>\lambda$ we have $K(\mathbf{x})>\lambda$. Let $\Sigma_I=\{\mathbf{x}\in\mathbb{R}^N : 2^{-I}\cdot\mathbf{x}\in\Sigma\}$. Then $\big|\Sigma_I\big| = 2^{+\sum_{j=1}^n i_jQ_j}\big|\Sigma\big|$. We have

$$\begin{aligned}\Big\{\mathbf{x}\in\mathbb{R}^N : \Big(\exists I\in\Gamma_{\mathbb{Z}}(\mathbf{E})\Big)\Big(2^{-I}\cdot\mathbf{x}\in\Sigma \text{ and } 2^{-\sum_{j=1}^n i_jQ_j}>\lambda\Big)\Big\}& \\ \subset \{\mathbf{x}\in\mathbb{R}^N : K(\mathbf{x})>\lambda\},&\end{aligned}$$

so

$$\begin{aligned}&\Big|\Big\{\mathbf{x}\in\mathbb{R}^N : K(\mathbf{x})>\lambda\Big\}\Big| \\ &\qquad \geq \Big|\Big\{\mathbf{x}\in\mathbb{R}^N : \Big(\exists I\in\Gamma_{\mathbb{Z}}(\mathbf{E})\Big)\Big(2^{-I\cdot\mathbf{x}}\in\Sigma \text{ and } 2^{-\sum_{j=1}^n i_jQ_j}>\lambda\Big)\Big\}\Big| \\ &\qquad \gtrsim \sum_{I\in\Gamma_{\mathbb{Z}}(\mathbf{E},\lambda)} |\Sigma_I| \gtrsim \lambda^{-1}\#\Big(\Gamma_{\mathbb{Z}}(\mathbf{E},\lambda)\Big)\end{aligned}$$

where $\Gamma_{\mathbb{Z}}(\mathbf{E},\lambda)$ is the set of n-tuples $I=(i_1,\ldots,i_n)\in\Gamma_{\mathbb{Z}}(\mathbf{E})$ such that $\lambda< 2^{-\sum_{j=1}^n i_jQ_j}\leq C\lambda$ with C a large constant. The cone $\Gamma(\mathbf{E})=\Big\{t=(t_1,\ldots,t_n) : e(j,k)t_k<t_j<0\Big\}$ is open since $\mathbf{E}$ has rank n, and the number of lattice points $(i_1,\ldots,i_n)$ in this cone satisfying

$$\log(\lambda)<\sum_{j=1}^n Q_ji_j<\log(\lambda)+1$$

is of the order of $\big(\log(\lambda)\big)^{n-1}$. This shows that $\#\Big(\Gamma_{\mathbb{Z}}(\mathbf{E},\lambda)\Big)\gtrsim\big(\log(\lambda)\big)^{n-1}$ which completes the proof of part (b). □

CHAPTER 8

Convolution operators on homogeneous nilpotent Lie groups

We now turn to the study of convolution operators $f \to f * \mathcal{K}$ with $\mathcal{K} \in \mathcal{P}_0(\mathbf{E})$ and the convolution on a homogeneous nilpotent Lie group G. The underlying manifold of G is $\mathbb{R}^N$ for some integer $N \geq 1$, and there is a one-parameter group of *automorphic dilations*, i.e., of automorphisms $\delta_r : G \to G$ which, in appropriate coordinates on G, take the form

$$\delta_r(x_1, \ldots, x_N) = (r^{d_1}x_1, \ldots, r^{d_N}x_N), \tag{8.1}$$

with $d_i > 0$ for every i. The ordering of the coordinates and the parametrization by r can be chosen so that

$$1 \leq d_1 \leq d_2 \leq \cdots \leq d_N. \tag{8.2}$$

In these coordinates, the product on G has the following form. For $\mathbf{x} = (x_1, \ldots, x_N)$, $\mathbf{y} = (y_1, \ldots, y_N)$,

$$\mathbf{xy} = \big(x_1 + y_1, x_2 + y_2 + M_2(\mathbf{x}, \mathbf{y}), \ldots, x_N + y_N + M_N(\mathbf{x}, \mathbf{y})\big), \tag{8.3}$$

where $M_2, \ldots, M_N$ are polynomials which vanish for $\mathbf{x}$ or $\mathbf{y}$ equal to 0 and are such that

$$M_l(\delta_r\mathbf{x}, \delta_r\mathbf{y}) = \delta_r M_l(\mathbf{x}, \mathbf{y}). \tag{8.4}$$

In particular, each M_l only depends on the variables x_m, y_m for which $d_m < d_l$. Moreover, the Haar measure for G is Lebesgue measure on $\mathbb{R}^N$.

If $f, g \in L^1(G)$, the convolution $f * g$ is defined by

$$f * g(\mathbf{x}) = \int_{G \cong R^N} f(\mathbf{xy}^{-1})g(\mathbf{y})\, d\mathbf{y} = \int_{G \cong \mathbb{R}^N} f(\mathbf{y})g(\mathbf{y}^{-1}\mathbf{x})\, d\mathbf{y}.$$

8.1. Convolution of scaled bump functions; compatibility of dilations and convolution

If $f \in L^1(G)$ and $\lambda = (\lambda_1, \ldots, \lambda_N)$ with each $\lambda_j > 0$, consider the N-parameter family of dilations of f by λ given by

$$f_\lambda(x_1, \ldots, x_N) = \Big[\prod_{j=1}^N \lambda_j^{\ d_j}\Big]\, f(\lambda_1^{-d_1}x_1, \ldots, \lambda_N^{-d_N}x_N), \tag{8.5}$$

If φ, ψ are normalized bump functions supported in the unit ball and if $\lambda = (\lambda_1, \ldots, \lambda_N)$ and $\mu = (\mu_1, \ldots, \mu_N)$, it is easy to check that for Euclidean convolution $\varphi_\lambda * \psi_\mu = \theta_\nu$ where $\theta \in \mathcal{C}_0^\infty(\mathbb{R}^N)$ is normalized relative to φ and ψ, and $\nu_j = \max\{\lambda_j, \mu_j\} = \lambda_j \vee \mu_j$. This fact is very useful in studying the convolution of two sums of dilates of bump functions. However, on a more general nilpotent Lie group, this need not be true unless we suitably restrict the N-tuples $\lambda = (\lambda_1, \ldots, \lambda_N)$ and $\mu = (\mu_1, \ldots, \mu_N)$. Let

$$E_N = \Big\{\lambda = (\lambda_1, \ldots, \lambda_N) \in (0, \infty)^N : \lambda_1 \leq \lambda_2 \leq \cdots \leq \lambda_N\Big\}. \tag{8.6}$$

We say that a mapping $(x_1, \ldots, x_N) \to (\lambda_1^{-d_1} x_1, \ldots, \lambda_N^{-d_N} x_N)$ is *compatible with the group structure of* G if $\lambda \in E_N$. The significance of this notion comes from the following result, which is Lemma 6.17 in [**NRSW12**].

LEMMA 8.1. *Let $\varphi, \psi \in C_0^\infty(G)$ be supported in $\mathbb{B}(1)$. If $\lambda, \mu \in E_N$ there exists $\theta \in C_0^\infty(G)$ so that $\varphi_\lambda * \psi_\mu = \theta_\nu$ where $\nu_j = \lambda_j \vee \mu_j = \max\{\lambda_j, \mu_j\}$. Moreover, θ is normalized relative to φ and ψ: there are constants $\rho_0 \geq 1$ and $C_m > 0$ independent of φ and ψ so that θ is supported in $\mathbb{B}(\rho_0)$ and* $\sup_{|\gamma| \leq m} ||\partial^\gamma \theta||_\infty \leq C_m \sup_{|\gamma| \leq m} ||\partial^\gamma \varphi||_\infty \sup_{|\gamma| \leq m} ||\partial^\gamma \psi||_\infty$.

It will be useful for our purposes to have a formulation of Lemma 8.1 for dilations with fewer parameters and with the N variables split into n blocks according to a decomposition $G \cong \mathbb{R}^N = \mathbb{R}^{C_1} \oplus \cdots \oplus \mathbb{R}^{C_n}$. For reasons of compatibility with the group structure, we now require that $C_1, \ldots, C_n \subset \{1, \ldots, N\}$ be sets of consecutive integers, indexed according to the natural ordering:

$$C_1 = \{1, \ldots, l_1\}, \quad C_2 = \{l_1 + 1, \ldots, l_2\}, \quad \ldots \quad C_n = \{l_{n-1} + 1, \ldots, N\}. \tag{8.7}$$

This implies that each $G_j \cong \mathbb{R}^{C_j} \oplus \cdots \oplus \mathbb{R}^{C_n}$ is a subgroup of G. Formula (8.3) remains true if the individual variables $x_l, y_l \in \mathbb{R}$ are replaced by the blocks of variables $\mathbf{x}_j, \mathbf{y}_j \in \mathbb{R}^{C_j}$ and the scalar-valued functions M_l by the $\mathbb{R}^{C_j}$-valued functions $\mathbf{M}_j = (M_{l_{j-1}+1}, \ldots, M_{l_j})$. Notice that the functions $\mathbf{M}_j$ still satisfy (8.4).[1]

Let $\mathbf{E} = \{e(j,k)\}$ be an $n \times n$ matrix satisfying (2.10). Associated to $\mathbf{E}$ is the cone of lattice points

$$\Gamma_{\mathbb{Z}}(\mathbf{E}) = \Big\{I = (i_1, \ldots, i_n) \in \mathbb{Z}^n : e(j,k) i_k \leq i_j < 0, \ 1 \leq j, k \leq n\Big\}$$

and we consider dilations

$$[f]_I(\mathbf{x}) = 2^{-\sum_{j=1}^n i_j Q_j} f(2^{-i_1} \cdot \mathbf{x}_1, \ldots, 2^{-i_n} \cdot \mathbf{x}_n).$$

Note that $(2^{i_1}, \ldots, 2^{i_n})$ is an n-tuple while elements of E_N defined in equation (8.6) are N-tuples. However $[f]_I = f_{\lambda(I)}$ where $\lambda(I)$ is obtained from I by setting $\lambda_l = 2^{i_k}$ if $l \in C_k$; that is

$$\lambda(I) = \big(\overbrace{2^{i_1}, \ldots, 2^{i_1}}^{C_1},\ \overbrace{2^{i_2}, \ldots, 2^{i_2}}^{C_2},\ \ldots,\ \overbrace{2^{i_n}, \ldots, 2^{i_n}}^{C_n}\big). \tag{8.8}$$

We want to impose conditions on the matrix $\mathbf{E}$, in addition to the basic hypothesis (2.10), which guarantees that if $I \in \Gamma_{\mathbb{Z}}(\mathbf{E})$, then $\lambda(I) \in E_N$.

[1] In what follows below, after choosing marked partitions, we may have situations where we no longer have sequences of consecutive integers. However this is all right if we can refine to a situation of consecutive integers.

DEFINITION 8.2. An $n \times n$ matrix $\mathbf{E} = \{e(j,k)\}$ is doubly monotone if each row is weakly increasing from left to right, and each column is weakly decreasing from top to bottom; *i.e.*

$$\begin{aligned} e(j,k) &\leq e(j,k+1) && \text{for } 1 \leq j \leq n \text{ and } 1 \leq k < n, \\ e(j,k) &\geq e(j+1,k) && \text{for } 1 \leq j < n \text{ and } 1 \leq k \leq n. \end{aligned}$$

PROPOSITION 8.3. Let $\mathbf{E} = \{e(j,k)\}$ be an $n \times n$ matrix satisfying the basic hypothesis (2.10).

(a) The matrix $\mathbf{E}$ is doubly monotone if and only if $e(j+1,j) \leq 1$ for $1 \leq j \leq n-1$.

(b) If the matrix $\mathbf{E}$ is doubly monotone and if $I \in \Gamma(\mathbf{E})$ then $\lambda(I) \in E_N$.

PROOF. If $\mathbf{E}$ is doubly monotone and each row is weakly increasing, then since $e(j+1,j+1) = 1$ we certainly have $e(j+1,j) \leq 1$. Conversely, suppose that $e(k+1,k) \leq 1$ for $1 \leq k \leq n-1$. Then by (2.10) we have

$$\begin{aligned} e(j,k) &\leq e(j,k+1)e(k+1,k) \leq e(j,k+1), \\ e(j+1,k) &\leq e(j+1,j)e(j,k) \leq e(j,k), \end{aligned}$$

so $\mathbf{E}$ is doubly monotone.

Next if $\mathbf{E}$ is doubly monotone and $I = (i_1, \ldots, i_n) \in \Gamma(\mathbf{E})$, it follows from the definition of $\lambda(I)$ in equation (8.8) that we only need to show that $i_j \leq i_{j+1}$ for $1 \leq j \leq n-1$. However, since $e(j+1,j) \leq 1$ and $i_j < 0$, we have $i_j \leq e(j+1,j)i_j \leq i_{j+1}$ since $I \in \Gamma_{\mathbb{Z}}(\mathbf{E})$. □

COROLLARY 8.4. Suppose $\mathbf{E}$ is doubly monotone, and let $I = (i_1, \ldots, i_n)$, $J = (j_1, \ldots, j_n) \in \Gamma_{\mathbb{Z}}(\mathbf{E})$. If $\varphi, \psi \in \mathcal{C}_0^\infty(\mathbb{R}^n)$ are supported in the unit ball, then $[\varphi]_I * [\psi]_J = [\theta]_K$ where $\theta \in \mathcal{C}_0^\infty(\mathbb{R}^N)$ is normalized relative to φ and ψ, and $K = (k_1, \ldots, k_n) \in \Gamma_{\mathbb{Z}}(\mathbf{E})$ with $k_m = \max\{i_m, j_m\}$ for $1 \leq m \leq n$.

Recall that $S \in \mathcal{S}'(\mathbf{E})$ means that $\widehat{E}_S \cap \mathbb{B}(1)^c \neq \emptyset$. If

$$S = \big((I_1,k_1); \ldots; (I_s,k_s)\big) \in \mathcal{S}'(\mathbf{E}),$$

the $s \times s$ matrix $\mathbf{E}_S = \{e_S(r,p)\}$ was defined in Lemma 6.2. We observe that if we order the sets $\{I_1, \ldots, I_s\}$ so that $k_1 \leq k_2 \leq \cdots \leq k_s$ and if $\mathbf{E}$ is doubly monotone, then the same is true of the matrix $\mathbf{E}_S$.

LEMMA 8.5. *Let* $S = \big((I_1,k_1); \ldots; (I_s,k_s)\big) \in \mathcal{S}'(\mathbf{E})$ *with* $k_1 < k_2 < \cdots < k_s$, *and let* $\mathbf{E}_S = \{e_S(r,p)\}$ *be the* $s \times s$ *matrix whose existence is established in Lemma 6.2. If the matrix* $\mathbf{E}$ *is doubly monotone, then the matrix* $\mathbf{E}_S$ *is doubly monotone.*

PROOF. From the construction of $\mathbf{E}_S$ we have

$$e_S(r+1,r) \leq \tau_S(r+1,r) = \min_{j \in I_r} \frac{e(k_{r+1},j)}{e(k_r,j)} \leq e(k_{r+1},k_r).$$

If $\{e(j,k)\}$ is doubly monotone, $e(k_{r+1},k_r) \leq 1$ because of our ordering of the sets $\{I_1, \ldots, I_s\}$, and it follows that $e_S(r+1,r) \leq 1$. Since $\mathbf{E}_S$ satisfies the basic hypothesis (2.10), it then follows from part (a) of Proposition 8.3 that it is also doubly monotone. □

Continue to fix $S = \big((I_1, k_1); \ldots ; (I_s, k_s)\big) \in \mathcal{S}'(\mathbf{E})$ with $k_1 < k_2 < \cdots < k_s$. In Definition 3.12 we introduced the family of dilations

$$2^{-J} \hat{\cdot}_S \mathbf{t} = \big(2^{-j_1} \hat{\cdot}_S \mathbf{t}_{I_1}, \ldots, 2^{-j_r} \hat{\cdot}_S \mathbf{t}_{I_r}, \ldots, 2^{-j_s} \hat{\cdot}_S \mathbf{t}_{I_s}\big)$$

where

$$2^{-j_r} \hat{\cdot}_S \mathbf{t}_{I_r} = \Big\{2^{-j_r/e(k_r,j)} \cdot \mathbf{t}_j : j \in I_r\Big\} = \Big\{2^{-j_r \tau_S(j,k_r)} \cdot \mathbf{t}_j : j \in I_r\Big\}.$$

Recall that $\Gamma_{\mathbb{Z}}(\mathbf{E}_S) = \big\{(j_1, \ldots, j_s) \in \mathbb{Z}^s : e_S(p, r) j_r \le j_p < 0\big\}$.

LEMMA 8.6. *Suppose that* $\mathbf{E}$ *is doubly monotone. If* $J = (j_1, \ldots, j_s) \in \Gamma_{\mathbb{Z}}(\mathbf{E}_S)$ *then the mapping* $\mathbf{t} \to 2^{-J} \hat{\cdot}_S \mathbf{t}$ *is compatible with the group structure.*

PROOF. Let $l, m \in \{1, \ldots, n\}$ with $l \le m$ and $l \in I_r$, $m \in I_p$. According to Proposition 8.3, it suffices to show that $j_r e(k_r, l)^{-1} \le j_p e(k_p, m)^{-1}$. If $J = (j_1, \ldots, j_s) \in \Gamma_{\mathbb{Z}}(\mathbf{E}_S)$ then $e_S(p, r) j_r \le j_p$. But then since $j_r < 0$,

$$\frac{j_p}{j_r} \le e_S(p, r) \le \tau_S(k_p, k_r) = \min_{j \in I_r} \frac{e(k_p, j)}{e(k_r, j)}$$
$$\le \frac{e(k_p, l)}{e(k_r, l)} \le \frac{e(k_p, m) e(m, l)}{e(k_r, l)} \le \frac{e(k_p, m)}{e(k_r, l)}$$

where the last inequality follows since $l \le m$ and the matrix $\mathbf{E}$ is (weakly) increasing across each row. This means that $j_r e(k_r, l)^{-1} \le j_p e(k_p, m)^{-1}$, completing the proof. □

8.2. Automorphic flag kernels and L^p-boundedness of convolution operators

In this section we discuss L^p-boundedness of operators

$$T_{\mathcal{K}} f = f * \mathcal{K},$$

given by convolution (relative to some homogeneous group structure on $\mathbb{R}^N$) with a kernel $\mathcal{K} \in \mathcal{P}(\mathbf{E})$. The matrix $\mathbf{E}$ is assumed to satisfy the basic hypotheses (2.10).

The reader should be aware that there are two possible interpretations of the symbol $f * \mathcal{K}$, depending on whether the underlying group is the standard abelian $\mathbb{R}^n$ or some non-abelian group G. Whenever clarification is required, we will write $T^G_{\mathcal{K}}$ to specify the group. For ordinary convolution on $\mathbb{R}^N$, it can be easily verified that a kernel in $\mathcal{P}(\mathbf{E})$ is a *product kernel*, as defined in Section 2 of [**NRS01**]. In fact, the differential inequalities of a product kernel follow from the inequalities $N_j(\mathbf{x}) \ge n_j(\mathbf{x}_j)$, $j = 1, \ldots, n$, and the cancellation conditions are the same as in Definition 2.2 (b). By [**NRS01**], we have the following theorem.

THEOREM 8.7. *Let* δ_λ *be any family of dilations on* $\mathbb{R}^N$ *and* $\mathbb{R}^N = \mathbb{R}^{C_1} \oplus \cdots \oplus \mathbb{R}^{C_n}$ *be any decomposition of* $\mathbb{R}^N$ *into homogeneous subspace. Suppose that the matrix* $\mathbf{E}$ *satisfies the basic hypotheses. Then, for every* $\mathcal{K} \in \mathcal{P}(\mathbf{E})$, *the operator* $T_{\mathcal{K}}$ *is bounded on* L^p *for* $1 < p < \infty$.

Considering now a general homogeneous group law as in (8.3), we must be more specific on all the ingredients needed to define a class $\mathcal{P}(\mathbf{E})$.

We fix a decomposition $\mathbb{R}^N = \mathbb{R}^{C_1} \oplus \cdots \oplus \mathbb{R}^{C_n}$ where the C_j are as in (8.7). In the notation of [**NRS01**], the subgroups $G_j = \mathbb{R}^{C_j} \oplus \cdots \oplus \mathbb{R}^{C_n}$ form the *standard flag*

$$\mathcal{F} : \qquad (0) \subset G_n \subset G_{n-1} \subset \cdots \subset G_2 \subset G_1 = G.$$

On each $\mathbb{R}^{C_j}$ we also fix a homogeneous norm n_j for the automorphic dilations δ_λ in (8.1). A distribution $\mathcal{K}$ is an *automorphic flag kernel* on the standard flag if it satisfies the differential inequalities

$$\left|\partial^\gamma K(\mathbf{x})\right| \leq C_\gamma \prod_{j=1}^{n} \left[n_1(\mathbf{x}_1) + \cdots + n_j(\mathbf{x}_j)\right]^{-Q_j - [\![\gamma_j]\!]}$$

and the cancellation conditions in Definition 2.2 (b).

The differential inequalities for the *automorphic flag multiplier* $m = \widehat{\mathcal{K}}$ are

$$\left|\partial^\gamma m(\boldsymbol{\xi})\right| \leq C_\gamma \prod_{j=1}^{n} \left[n_j(\boldsymbol{\xi}_j) + \cdots + n_n(\boldsymbol{\xi}_n)\right]^{-[\![\gamma_j]\!]}.$$

(See Definition 2.3.2 in [**NRS01**] for details.)

We show that, if $\mathbf{E}$ is doubly monotone, a distribution $\mathcal{K} \in \mathcal{P}_0(\mathbf{E})$ is always an automorphic flag kernel for a standard flag.

PROPOSITION 8.8. If $\mathcal{K} \in \mathcal{P}_0(\mathbf{E})$ where the class of distributions is defined by a doubly monotone matrix, then $\mathcal{K}$ is an automorphic flag kernel on the standard flag.

PROOF. Let $\mathcal{K} \in \mathcal{P}_0(\mathbf{E})$ where the class of distributions is defined by a doubly monotone matrix $\mathbf{E} = \{e(j,k)\}$ and let $m = \widehat{\mathcal{K}}$. If $\boldsymbol{\xi} \in \mathbb{B}(1)^c$ then $n_k(\boldsymbol{\xi}_k) \geq 1$ for at least one index k and so $N_j(\boldsymbol{\xi}) = \sum_{k=1}^n n_k(\boldsymbol{\xi}_k)^{1/e(k,j)} \geq 1$. Let

$$\begin{aligned} A_j(\boldsymbol{\xi}) &= \Big\{k \in \{j, \ldots, n\} : n_k(\boldsymbol{\xi}_k) \geq 1\Big\}, \\ B_j(\boldsymbol{\xi}) &= \Big\{k \in \{j, \ldots, n\} : n_k(\boldsymbol{\xi}_k) < 1\Big\} \end{aligned}$$

Then

$$\widehat{N}_j(\boldsymbol{\xi}) \geq \sum_{k \in A_j(\boldsymbol{\xi})} n_k(\boldsymbol{\xi}_k)^{1/e(k,j)} \geq \sum_{k \in A_j(\boldsymbol{\xi})} n_k(\boldsymbol{\xi}_k)$$

since for $k \geq j$ we have $e(k,j) \leq 1$. On the other hand,

$$\sum_{k \in B_j(\boldsymbol{\xi})} n_k(\boldsymbol{\xi}_k)^{1/e(k,j)} \leq n \leq n\, N_j(\boldsymbol{\xi})$$

and adding the two inequalities shows that $N_j(\boldsymbol{\xi}) \geq (n+1)^{-1} \sum_{k=j}^n n_k(\boldsymbol{\xi}_k)$. Thus if $\gamma \in \mathbb{N}^{C_1} \times \cdots \times \mathbb{N}^{C_n}$ we have

$$\left|\partial^\gamma m(\boldsymbol{\xi})\right| \leq C_\gamma \prod_{j=1}^{n} \left(1 + \widehat{N}_j(\boldsymbol{\xi})\right)^{-[\![\gamma_j]\!]} \leq C_\gamma \prod_{j=1}^{n} \left[n_j(\boldsymbol{\xi}_j) + \cdots + n_n(\boldsymbol{\xi}_n)\right]^{-[\![\gamma_j]\!]}.$$

□

THEOREM 8.9. *Let G be a homogeneous nilpotent Lie group, endowed with automorphic dilations δ_λ, $G \cong \mathbb{R}^N = \mathbb{R}^{C_1} \oplus \cdots \oplus \mathbb{R}^{C_n}$, the C_j being as in (8.7). If $\mathbf{E}$ is a doubly monotone matrix satisfying the basic hypothesis, and if $\mathcal{K} \in \mathcal{P}_0(\mathbf{E})$, then the operator $T_{\mathcal{K}}[\varphi] = \varphi * \mathcal{K}$, defined initially for $\varphi \in \mathcal{S}(\mathbb{R}^N)$, extends uniquely to a bounded operator on $L^p(G)$ for $1 < p < \infty$. If the rank of $\mathbf{E}$ is greater than one, the same holds for $\mathcal{K} \in \mathcal{P}(\mathbf{E})$.*

PROOF. If $\mathcal{K} \in \mathcal{P}_0(\mathbf{E})$, it follows from Proposition 8.8 that $\mathcal{K}$ is an automorphic flag kernel on the standard flag. The result now follows from Theorem 8.14 in [**NRSW12**]. Assume now that the rank of $\mathbf{E}$ is greater than 1 and $\mathcal{K} \in \mathcal{P}(\mathbf{E})$. By Corollary 7.4, we can write $\mathcal{K} = \mathcal{K}_0 + K_\infty$ where $\mathcal{K}_0 \in \mathcal{P}_0(\mathbf{E})$ and $K_\infty \in L^1(\mathbb{R}^N)$. The operator $f \to f * K_\infty$ is bounded on $L^p(G)$ for $1 \leq p \leq \infty$. $\square$

CHAPTER 9

Composition of operators

If $\mathcal{K} \in \mathcal{P}_0(\mathbf{E})$ let $T_{\mathcal{K}}[f] = f * \mathcal{K}$ be the corresponding left-invariant convolution operator on the homogeneous nilpotent Lie group G. $T_{\mathcal{K}}$ is initially defined on the Schwartz space $\mathcal{S}(\mathbb{R}^N)$. We are interested in the composition of two operators $T_{\mathcal{K}}, T_{\mathcal{L}}$ with $\mathcal{K}, \mathcal{L} \in \mathcal{P}_0(\mathbf{E})$. Formally $T_{\mathcal{L}} \circ T_{\mathcal{K}}[f] = T_{\mathcal{K}}[f] * \mathcal{L} = (f * \mathcal{K}) * \mathcal{L} = f * (\mathcal{K} * \mathcal{L}) = T_{\mathcal{K} * \mathcal{L}}$.

To give a concrete meaning to these equalities is easy if convolution is relative to the (abelian) additive structure of $\mathbb{R}^N$, because we can refer to the Fourier multipliers $m = \widehat{\mathcal{K}}$, $m' = \widehat{\mathcal{L}}$. For $f \in \mathcal{S}(\mathbb{R}^N)$, $T_{\mathcal{L}} \circ T_{\mathcal{K}}[f] = \mathcal{F}^{-1}(mm'\widehat{f})$, where $\mathcal{F}$ denotes the Fourier transform, and the convolution kernel of $T_{\mathcal{L}} \circ T_{\mathcal{K}}$ is $\mathcal{F}^{-1}(mm') = \mathcal{K} * \mathcal{L}$. In fact, since $\mathcal{M}_\infty(\mathbf{E})$ is closed under pointwise product, we can conclude as follows.

THEOREM 9.1. *Let $\mathbf{E}$ be a matrix satisfying the basic hypotheses and let $\mathcal{K}, \mathcal{L} \in \mathcal{P}_0(\mathbf{E})$. With respect to ordinary convolution on $\mathbb{R}^N$, then $\mathcal{K} * \mathcal{L} \in \mathcal{P}_0(\mathbf{E})$.*

Suppose now that $\mathbb{R}^N \cong G$ is a noncommutative homogeneous group. It is still easy to verify that $\mathcal{K} * \mathcal{L}$ is a well-defined distribution. In fact,

$$\mathcal{K} = \mathcal{K}_0 + \varphi_0, \qquad \mathcal{L} = \mathcal{L}_0 + \psi_0,$$

where $\mathcal{K}_0, \mathcal{L}_0$ have compact support and φ_0, ψ_0 are Schwartz functions. Hence

$$\mathcal{K} * \mathcal{L} = \mathcal{K}_0 * \mathcal{L}_0 + \mathcal{K}_0 * \psi_0 + \varphi_0 * \mathcal{L}_0 + \varphi_0 * \psi_0,$$

where each summand is well defined. However, it is not guaranteed that the convolution $\mathcal{K} * \mathcal{L}$ is in the same class.

The main result of this chapter is that, under the assumptions introduced in Section 8, $\mathcal{K} * \mathcal{L} \in \mathcal{P}_0(\mathbf{E})$, and so $\mathcal{P}_0(\mathbf{E})$ is an algebra under convolution on the group G. We recall the assumptions:

(i) the underlying dilations on $\mathbb{R}^N$ are automorphisms of G;
(ii) the subspaces $\mathbb{R}^{C_j}$ are as in (8.7);
(iii) the matrix $\mathbf{E}$ is doubly monotone.

THEOREM 9.2. *Suppose that conditions* (i)-(iii) *above are satisfied. If $\mathcal{K}, \mathcal{L} \in \mathcal{P}_0(\mathbf{E})$ then $\mathcal{K} * \mathcal{L} \in \mathcal{P}_0(\mathbf{E})$.*

The following outline provides an informal *road map* for the long proof which begins in Section 9.1. Let $\mathcal{K}, \mathcal{L} \in \mathcal{P}_0(\mathbf{E})$.

(A) According to Theorem 6.14,

$$\mathcal{K} = \varphi_0 + \sum_{S \in \mathcal{S}'(\mathbf{E})} \mathcal{K}_S \quad \text{and} \quad \mathcal{L} = \psi_0 + \sum_{T \in \mathcal{S}'(\mathbf{E})} \mathcal{L}_T$$

where $\varphi_0, \psi_0 \in \mathcal{S}(\mathbb{R}^N)$ and

$$\mathcal{K}_S = \sum_{I \in \Gamma_{\mathbb{Z}}(\mathbf{E}_S)} [\varphi_S^I]_I \quad \text{and} \quad \mathcal{L}_T = \sum_{J \in \Gamma_{\mathbb{Z}}(\mathbf{E}_T)} [\psi_T^J]_J.$$

Here $\{\varphi_S^I\}$ and $\{\psi_T^J\}$ are uniformly bounded families of functions in $\mathcal{C}_0^\infty(\mathbb{R}^N)$, and

$$\Gamma_{\mathbb{Z}}(\mathbf{E}_S) = \Big\{I = (i_1, \ldots, i_s) \in \mathbb{Z}^s : \tau_S(k_p, k_r) i_r \leq i_p < 0,\ 1 \leq r, p \leq s\Big\},$$
$$\Gamma_{\mathbb{Z}}(\mathbf{E}_T) = \Big\{J = (j_1, \ldots, j_q) \in \mathbb{Z}^q : \tau_T(\ell_b, \ell_a) j_a \leq j_b < 0,\ 1 \leq a, b \leq q\Big\}.$$

Then the problem of studying the composition $T_{\mathcal{L}} \circ T_{\mathcal{K}}$ reduces essentially to the study of the finite number of compositions $\{T_{\mathcal{L}_T} \circ T_{\mathcal{K}_S}\}$ for $S, T \in \mathcal{S}'(\mathbf{E})$. This reduction is discussed in Section 9.1.

(B) We fix marked partitions

$$S = \big((I_1, k_1); \ldots; (I_s, k_s)\big) \qquad \text{and} \quad T = \big((J_1, \ell_1); \ldots; (J_q, \ell_q)\big)$$

in $\mathcal{S}'(\mathbf{E})$, and we simplify notation by writing $\varphi_S^I = \varphi^I$ and $\psi_T^J = \psi^J$. Thus we need to study $T_{\mathcal{L}_T} \circ T_{\mathcal{K}_S}$ where

$$\mathcal{K}_S = \sum_{I \in \Gamma_{\mathbb{Z}}(\mathbf{E}_S)} [\varphi^I]_I = \lim_{F_1 \nearrow \Gamma_{\mathbb{Z}}(\mathbf{E}_S)} \sum_{I \in F_1} [\varphi^I]_I = \lim_{F \nearrow \Gamma_{\mathbb{Z}}(\mathbf{E}_S)} \mathcal{K}_{F_1, S},$$
$$\mathcal{L}_T = \sum_{J \in \Gamma_{\mathbb{Z}}(\mathbf{E}_T)} [\psi^J]_J = \lim_{F_2 \nearrow \Gamma_{\mathbb{Z}}(\mathbf{E}_T)} \sum_{J \in F_2} [\psi^J]_J = \lim_{F_2 \nearrow \Gamma_{\mathbb{Z}}(\mathbf{E}_S)} \mathcal{L}_{F_2, T}.$$

Here F_1, F_2 are finite sets of indices, and the limits exist in the sense of distributions.

(C) Recall from Proposition 6.1 that, since $S, T \in \mathcal{S}'(\mathbf{E})$, $\Gamma_{\mathbb{Z}}(\mathbf{E}_S)$ and $\Gamma_{\mathbb{Z}}(\mathbf{E}_T)$ have nonempty interiors. By Lemma 6.2 in Section 6.1, there are new matrices $\mathbf{E}_S = \{e_S(p,r)\}$ and $\mathbf{E}_T = \{e_T(b,a)\}$ which satisfy the basic hypothesis such that $e_S(p,r) \leq \tau_S(k_p, k_r)$, $e_T(b,a) \leq \tau_T(\ell_b, \ell_a)$.

If $F_1 \subset \Gamma_{\mathbb{Z}}(\mathbf{E}_S)$ and $F_2 \subset \Gamma_{\mathbb{Z}}(\mathbf{E}_T)$ are any two finite sets of indices then $T_{\mathcal{L}_{F_2,T}} \circ T_{\mathcal{K}_{F_1,S}} = T_{\mathcal{H}_{F_1,F_2,S,T}}$ where

$$\mathcal{H}_{F_1,F_2,S,T} = \mathcal{K}_{F_1,S} * \mathcal{L}_{F_2,T} = \sum_{I \in F_1} \sum_{J \in F_2} [\varphi^I]_I * [\psi^J]_J.$$

In Section 9.2 we see that it suffices to show that

$$\lim_{F_1 \nearrow \Gamma_{\mathbb{Z}}(\mathbf{E}_S)} \lim_{F_2 \nearrow \Gamma_{\mathbb{Z}}(\mathbf{E}_T)} \mathcal{H}_{F_1,F_2,S,T}$$

exists in the sense of distributions and belongs to $\mathcal{P}_0(\mathbf{E})$. More generally, for arbitrary finite subsets $F \subset \Gamma_{\mathbb{Z}}(\mathbf{E}_S) \times \Gamma_{\mathbb{Z}}(\mathbf{E}_T)$, we show that

(a) $\mathcal{H}_{F,S,T} = \sum_{(I,J) \in F} [\varphi^I]_I * [\psi^J]_J \in \mathcal{P}_0(\mathbf{E})$ with estimates independent of the set F;

(b) In the sense of distributions, $\lim_{F \nearrow \Gamma_{\mathbb{Z}}(\mathbf{E}_S) \times \Gamma_{\mathbb{Z}}(\mathbf{E}_T)} \mathcal{H}_{F,S,T}$ exists and belongs to $\mathcal{P}_0(\mathbf{E})$.

(D) In Section 9.3 we summarize the support, decay, and cancellation properties of the convolutions $[\varphi^I]_I * [\psi^J]_J$ that are needed in our analysis.

(E) In order to study $\mathcal{H}_{F,S,T}$ we write

$$\Gamma_{\mathbb{Z}}(\mathbf{E}_S) \times \Gamma_{\mathbb{Z}}(\mathbf{E}_T) = \bigcup_{W \in \mathcal{W}(S,T)} \Gamma_W(S,T)$$

where the index set $\mathcal{W}(S,T)$ is finite, and the sets $\{\Gamma_W(S,T) \subset \Gamma_{\mathbb{Z}}(\mathbf{E}_S) \times \Gamma_{\mathbb{Z}}(\mathbf{E}_T)\}$ are disjoint. This partition of $\Gamma_{\mathbb{Z}}(\mathbf{E}_S) \times \Gamma_{\mathbb{Z}}(\mathbf{E}_T)$ is somewhat complicated, but it has the property that if $(I,J) = ((i_1,\ldots,i_s),(j_1,\ldots,j_q)) \in \Gamma_W(S,T)$, then for $1 \le r \le s$ and $1 \le a \le q$, if $I_r \cap J_a \neq \emptyset$ the ratio i_r/j_a is suitably restricted. Having made the partition, it then suffices to show that for every $W \in \mathcal{W}(S,T)$, $\lim_{F \nearrow \Gamma_{W(S,T)}} \mathcal{H}_{F,S,T}$ converges in the sense of distributions and belongs to $\mathcal{P}_0(\mathbf{E})$. This partition is described in Section 9.4.

(F) Now fix $W \in \mathcal{W}(S,T)$, and let $F \subset \Gamma_W(S,T)$ be a finite set. We study of $\mathcal{H}_{F,S,T} = \sum_{(I,J)\in F} [\varphi^I]_I * [\psi^J]_J$.

(a) For each convolution $[\varphi^I]_I * [\psi^J]_J$ there is a normalized bump function $\theta^{I,J}$ and a set of dilations $M(I,J)$ so that $[\varphi^I]_I * [\psi^J]_J = [\theta^{I,J}]_{M(I,J)}$.

(b) There are s parameters $(i_1,\ldots,i_s)$ in I and q parameters $(j_1,\ldots,j_q)$ in J, and the function $[\theta^{I,J}]_{M(I,J)}$ is dilated by some collection of these dilations. However not all of these need appear in $M(I,J)$. We will say that a parameter i_r or j_a from $I \cup J$ is 'fixed' if it appears among the parameters in $M(I,J)$, and 'free' if it does not appear. These notions are defined precisely in Section 9.5.

(c) Let I_r^* be the indices in I_r which are fixed, and let J_a^* be the indices in J_a which are fixed. Some of these sets may be empty (if I_r or J_a consists entirely of free indices), and we suppose there are $\alpha \le s$ non-empty sets $I_{r_1}^*,\ldots,I_{r_\alpha}^*$ and $\beta \le q$ non-empty set $J_{a_1}^*,\ldots J_{a_\beta}^*$.

(d) The sets $I_{r_1}^*,\ldots,I_{r_\alpha}^*, J_{a_1}^*,\ldots J_{a_\beta}^*$ are disjoint and their union is $\{1,\ldots,n\}$. This leads to a new decomposition

$$\mathbb{R}^N = \bigoplus_{j=1}^{\alpha} \mathbb{R}^{I_{r_j}^*} \oplus \bigoplus_{k=1}^{\beta} \mathbb{R}^{J_{a_k}^*}.$$

This decomposition is discussed in Section 9.6.

(G) Associated to this last decomposition, we show in Section 9.7 that there is an $(\alpha+\beta)\times(\alpha+\beta)$ matrix $\mathbf{E}(S,T,W)$ with associated family of norms $N_1^\#,\ldots,N_{\alpha+\beta}^\#$ on $\mathbb{R}^N$. In Section 9.8 we show that the corresponding family of distributions $\mathcal{P}_0(\mathbf{E}(S,T,W))$ is contained in the original family $\mathcal{P}(\mathbf{E})$. Theorem 9.2 then follows if we can show that

$$\sum_{(I,J)\in\Gamma_W(S,T)} [\varphi^I]_I * [\psi^J]_J \in \mathcal{P}_0(\mathbf{E}(S,T,W)).$$

This is the heart of the matter, and we argue as follows.

a) Each convolution $[\varphi^I]_I * [\psi^J]_J = [\theta^{I,J}]_{M(I,J)}$ where $\theta^{I,J} \in \mathcal{C}_0^\infty(\mathbb{R}^N)$ is normalized relative to φ^I and ψ^J, and where $M(I,J)$ is an $\alpha+\beta$-tuple of integers coming from $\alpha+\beta$ of the integers in $I = (i_1,\ldots,i_s)$ and $J = (j_1,\ldots,j_q)$. Since each function φ^I and ψ^J has strong cancellation, the function $\theta^{I,J}$ will have appropriate decay in the free variables and weak cancellation in the fixed variables. (See Lemma 9.4 below.)

b) We let $\mathfrak{M}(S,T,W)$ denote the (infinite) set of $\alpha+\beta$-tuples $M = M(I,J)$ that arise in this way as the dilation for some convolution $[\varphi^I]_I * [\psi^J]_J$. For each $M \in \mathfrak{M}(S,T,W)$, we let $F(M)$ be the set of pairs $(I,J) \in \Gamma_W(S,T)$ such that $M(I,J) = M$. It follows that

$$\sum_{(I,J)\in\Gamma_W(S,T)} [\varphi^I]_I * [\psi^J]_J = \sum_{M\in\mathfrak{M}(S,T,W)} \Big[\sum_{(I,J)\in F(M)} \theta^{I,J}\Big]_M.$$

c) Because of the decay properties of $\theta^{I,J}$, for each $M \in \mathfrak{M}(S,T,W)$, the inner infinite sum $\sum_{(I,J)\in F(M)} \theta^{I,J}$ converges to a function $\Theta^M \in \mathcal{C}_0^\infty$, and the family of functions $\{\Theta^M\}$ is uniformly bounded and has weak cancellation relative to M. Thus

$$\sum_{(I,J)\in\Gamma_W(S,T)} [\varphi^I]_I * [\psi^J]_J = \sum_{M\in\mathfrak{M}(S,T,W)} [\Theta^M]_M.$$

Theorem 5.13 then shows that

$$\sum_{(I,J)\in\Gamma_W(S,T)} [\varphi^I]_I * [\psi^J]_J \in \mathcal{P}_0(\mathbf{E}(S,T,W)).$$

The details are provided in Sections 9.9 and 9.10.

9.1. A preliminary decomposition

Let $\mathcal{K},\mathcal{L} \in \mathcal{P}_0(\mathbf{E})$. According to Corollary 6.16 we can write

$$\mathcal{K} = \varphi_0 + \sum_{S\in\mathcal{S}'(\mathbf{E})} \mathcal{K}_S = \varphi_0 + \sum_{S\in\mathcal{S}'(\mathbf{E})} \sum_{I\in\Gamma_{\mathbb{Z}}(\mathbf{E}_S)} [\varphi_S^I]_I,$$
$$\mathcal{L} = \psi_0 + \sum_{T\in\mathcal{S}'(\mathbf{E})} \mathcal{L}_T = \psi_0 + \sum_{T\in\mathcal{S}'(\mathbf{E})} \sum_{J\in\Gamma_{\mathbb{Z}}(\mathbf{E}_T)} [\psi_T^J]_J,$$

where $\varphi_0,\psi_0 \in \mathcal{S}(\mathbb{R}^N)$ and $\{\varphi_S^I\}$ and $\{\psi_T^J\}$ are uniformly bounded families of functions in $\mathcal{C}_0^\infty(\mathbb{R}^N)$ supported in $\mathbb{B}(1)$. Each φ_S^I has cancellation in the variables $\{\mathbf{x}_{k_1},\dots,\mathbf{x}_{k_s}\}$ and each ψ_T^J has cancellation in the variables $\{\mathbf{x}_{l_1},\dots,\mathbf{x}_{l_q}\}$. We have

$$T_\mathcal{K} \circ T_\mathcal{L} = \varphi_0 * \mathcal{L} + \mathcal{K} * \psi_0 + \sum_{S,T\in\mathcal{S}'(\mathbf{E})} T_{\mathcal{K}_S} \circ T_{\mathcal{L}_T}.$$

Since the first two terms on the right-hand side are Schwartz functions, to prove Theorem 9.2 it suffices to show that each composition $T_{\mathcal{K}_S} \circ T_{\mathcal{L}_T}$ is a convolution operator $T_{\mathcal{H}_{S,T}}$ where $\mathcal{H}_{S,T} \in \mathcal{P}_0(\mathbf{E})$.

9.2. Reduction to the case of finite sets

It follows from Corollary 6.16 that

$$\mathcal{K}_S(\mathbf{x}) = \lim_{F\nearrow\Gamma_{\mathbb{Z}}(\mathbf{E}_S)} \sum_{I\in F} [\varphi_S^I]_I(\mathbf{x}), \qquad \mathcal{L}_T(\mathbf{x}) = \lim_{G\nearrow\Gamma_{\mathbb{Z}}(\mathbf{E}_T)} \sum_{J\in G} [\psi_T^J]_J(\mathbf{x})$$

where the limits are taken in the sense of distributions, F ranges over finite subsets of $\Gamma_{\mathbb{Z}}(\mathbf{E}_S)$, and G ranges over finite subsets of $\Gamma_{\mathbb{Z}}(\mathbf{E}_T)$. It follows that

$$T_{\mathcal{K}_S} \circ T_{\mathcal{L}_T}[f] = \lim_{F \nearrow \Gamma_{\mathbb{Z}}(\mathbf{E}_S)} \lim_{G \nearrow \Gamma_{\mathbb{Z}}(\mathbf{E}_T)} \Big[f * \sum_{I \in F} \sum_{J \in G} [\varphi_S^I]_I * [\psi_T^J]_J\Big].$$

Thus it will suffice to show that

$$\lim_{F \nearrow \Gamma_{\mathbb{Z}}(\mathbf{E}_S)} \lim_{G \nearrow \Gamma_{\mathbb{Z}}(\mathbf{E}_T)} \sum_{I \in F} \sum_{J \in G} [\varphi_S^I]_I * [\psi_T^J]_J \in \mathcal{P}_0(\mathbf{E})$$

where the limits are take in the sense of distributions. To do this, we will show that for any finite sets $F_1 \subset \Gamma_{\mathbb{Z}}(\mathbf{E}_S)$ and $F_2 \subset \Gamma_{\mathbb{Z}}(\mathbf{E}_T)$,

$$\mathcal{H}_{F_1,F_2} = \sum_{I \in F_1} \sum_{J \in F_2} [\varphi_S^I]_I * [\psi_T^J]_J \in \mathcal{P}_0(\mathbf{E})$$

with estimates that are independent of the choice of the finite subsets F and G, and

$$\mathcal{H}_{S,T} = \lim_{F_1 \nearrow \Gamma_{\mathbb{Z}}(\mathbf{E}_S)} \lim_{F_2 \nearrow \Gamma_{\mathbb{Z}}(\mathbf{E}_T)} \mathcal{H}_{F,G} \in \mathcal{P}_0(\mathbf{E})$$

Thus to prove Theroem 9.2, it will suffice establish the following result.

LEMMA 9.3. *For each $S, T \in \mathcal{S}'(\mathbf{E})$ there exists $\mathcal{H}_{S,T} \in \mathcal{P}_0(\mathbf{E})$ so that*

$$\mathcal{K}_S * \mathcal{L}_T = \sum_{(I,J) \in \Gamma_{\mathbb{Z}}(\mathbf{E}_S) \times \Gamma_{\mathbb{Z}}(\mathbf{E}_T)} [\varphi_S^I]_I * [\psi_T^J]_J = \mathcal{H}_{S,T} \tag{9.1}$$

where the series converges in the sense of distributions as the limit of sums over finite subsets of $\Gamma_{\mathbb{Z}}(\mathbf{E}_S) \times \Gamma_{\mathbb{Z}}(\mathbf{E}_T)$.

9.3. Properties of the convolution $[\varphi_S^I]_I * [\psi_T^J]_J$

In order to proceed further, we will need information about the support, decay, and cancellation properties of each of the convolutions $[\varphi_S^I]_I * [\psi_T^J]_J$ that appear in equation (9.1). The required estimates are given in Lemma 9.4 below. The statement essentially incorporates the results of Lemmas 6.17 and 6.18 and Corollary 6.19 in [**NRSW12**]. Since the setting now is somewhat more complicated, we include a sketch of the proof.

Suppose that G is a homogeneous nilpotent Lie group with underlying manifold $\mathbb{R}^N$, and that the one-parameter family of automorphic dilations on G is given by

$$\delta_r(\mathbf{x}) = (r^{d_1} x_1, \ldots, r^{d_N} x_N).$$

Let $*$ denote the convolution on G.

LEMMA 9.4. *Let $\varphi, \psi \in \mathcal{C}_0^\infty(\mathbb{R}^N)$ have support in the unit ball and put*

$$[\varphi]_\lambda(\mathbf{x}) = \Big[\prod_{m=1}^{N} \lambda_m^{-d_m}\Big] \varphi(\lambda_1^{-d_1} x_1, \ldots, \lambda_N^{-d_N} x_N),$$

$$[\psi]_\mu(\mathbf{x}) = \Big[\prod_{m=1}^{N} \mu_m^{-d_m}\Big] \psi(\mu_1^{-d_1} x_1, \ldots, \mu_N^{-d_N} x_N).$$

Assume that $0 < \lambda_1 \leq \lambda_2 \leq \cdots \leq \lambda_N$ and $0 < \mu_1 \leq \mu_2 \leq \cdots \leq \mu_N$.

(a) [Support properties] *There exists* $\theta \in \mathcal{C}_0^\infty(\mathbb{R}^N)$, *normalized with respect to* φ *and* ψ, *so that,*

$$[\varphi]_\lambda * [\psi]_\mu(\mathbf{x}) = \Big[\prod_{m=1}^{n} \nu_m^{-d_m}\Big]\theta(\nu_1^{-d_1}\mathbf{x}_1, \ldots, \nu_N^{-d_N}x_N) = [\theta]_\nu(\mathbf{x})$$

where $\nu_j = \lambda_j \vee \mu_j = \max\{\lambda_j, \mu_j\}$.

(b) [Cancellation and decay properties] *The estimates come in three parts, depending on whether there is cancellation in any of the first* $N-1$ *variables* $\{x_1, \ldots, x_{N-1}\}$, *or in the last variable* x_N.

(i) *Let* $C, D \subset \{1, \ldots, N-1\}$, *and suppose that* φ *has cancellation in the variables* x_j *for* $j \in C$, *and that* ψ *has cancellation in the variables* x_k *for* $k \in D$. *Put*

$$C_1 = \Big\{j \in \{1, \ldots, N-1\} : \varphi \textit{ has cancellation in } x_j \textit{ and } \lambda_j < \mu_j\Big\},$$

$$C_2 = \Big\{j \in \{1, \ldots, N-1\} : \varphi \textit{ has cancellation in } x_j \textit{ and } \lambda_j \geq \mu_j\Big\},$$

$$D_1 = \Big\{j \in \{1, \ldots, N-1\} : \psi \textit{ has cancellation in } x_j \textit{ and } \mu_j \leq \lambda_j\Big\},$$

$$D_2 = \Big\{j \in \{1, \ldots, N-1\} : \psi \textit{ has cancellation in } x_j \textit{ and } \mu_j > \lambda_j\Big\},$$

so that

$$\begin{aligned} C_1 \cup C_2 &= C, & C_1 \cap C_2 &= \emptyset, \\ D_1 \cup D_2 &= D, & D_1 \cap D_2 &= \emptyset, \\ C_1 \cap D_1 &= \emptyset, & C_2 \cap D_2 &= \emptyset. \end{aligned}$$

There exists $\epsilon > 0$ *depending only on the group structure so that* $[\varphi]_\lambda * [\psi]_\mu$ *can be written as a finite sum of terms of the form*

$$\begin{aligned} &\prod_{j\in C_1}\Big[\Big(\frac{\lambda_j}{\mu_j}\Big)^\epsilon + \Big(\frac{\lambda_j}{\lambda_{j+1}}\Big)^\epsilon\Big] \times \prod_{k\in D_1}\Big[\Big(\frac{\mu_k}{\lambda_k}\Big)^\epsilon + \Big(\frac{\mu_k}{\mu_{k+1}}\Big)^\epsilon\Big] \times \\ &\qquad \times \prod_{j\in C_2'}\Big(\frac{\lambda_j}{\lambda_{j+1}}\Big)^\epsilon \times \prod_{k\in D_2'}\Big(\frac{\mu_k}{\mu_{k+1}}\Big)^\epsilon [\Theta_{C_2'',D_2''}]_\nu \end{aligned} \tag{9.2}$$

where

$$\begin{aligned} C_2 &= C_2' \cup C_2'', & C_2' \cap C_2'' &= \emptyset, \\ D_2 &= D_2' \cup D_2'', & D_2' \cap D_2'' &= \emptyset. \end{aligned}$$

The functions $\{\Theta_{C_2'',D_2''}\} \subset \mathcal{C}_0^\infty(\mathbb{R}^N)$ *are normalized relative to* φ *and* ψ *and have cancellation in the variables* x_j *for* $j \in C_2'' \cup D_2''$.

(ii) *Suppose that* φ *also has cancellation in the variable* x_N. *If* $\lambda_N < \mu_N$, *there is an additional factor* $\big(\frac{\lambda_N}{\mu_N}\big)^\epsilon$ *in equation (9.2). If* $\lambda_N \geq \mu_N$ *then the function* $\Theta_{C_2'',D_2''}$ *in equation (9.2) also has cancellation in the variable* x_N.

(iii) *Suppose that ψ also has cancellation in the variable x_N. If $\lambda_N < \mu_N$, then the function $\Theta_{C_2'',D_2''}$ in equation (9.2) also has cancellation in the variable x_N. If $\lambda_N \geq \mu_N$ there is an additional factor $\left(\frac{\lambda_N}{\mu_N}\right)^\epsilon$ in equation (9.2).*

SKETCH OF THE PROOF.

Because of the monotonicity of the λ's and the μ's, assertion (a) follows immediately from Lemma 8.1, so the real content of the lemma is assertion (b). Let us first consider the case of Abelian convolution. Then there is no need to distinguish between the first $N-1$ variable and the last variable. If (say) φ has cancellation in the variable x_j then we can write $\varphi = \partial_{x_j}\phi$ where ϕ is normalized in terms of φ. Suppose that $\lambda_j < \mu_j$. Then moving the derivative ∂_{x_j} across the convolution, we have

$$\begin{aligned}[\varphi]_\lambda * [\psi]_\mu &= [\partial_{x_j}\phi]_\lambda * [\psi]_\mu = \lambda_j^{d_j}\partial_{x_j}([\phi]_\lambda) * [\psi]_\mu \\ &= \lambda_j^{d_j}[\phi]_\lambda * \partial_{x_j}([\psi]_\mu) = \left(\frac{\lambda_j}{\mu_j}\right)^{d_j}[\phi]_\lambda * [\partial_{x_j}\psi]_\mu = \left(\frac{\lambda_j}{\mu_j}\right)^{d_j}[\theta]_\nu\end{aligned}$$

where we have use part (a) to write $[\phi]_\lambda * [\partial_{x_j}\psi]_\mu = [\theta]_\nu$ where θ is normalized. On the other hand, if $\lambda_j \geq \mu_j$, we can put the derivative in front of the whole convolution, and we have

$$\begin{aligned}[\varphi]_\lambda * [\phi]_\mu &= [\partial_{x_j}\phi]_\lambda * [\phi]_\mu = \lambda_j^{d_j}\partial_{x_j}([\phi]_\lambda) * [\phi]_\mu \\ &= \lambda_j^{d_j}\partial_{x_j}\Big([\phi]_\lambda * [\psi]_\mu\Big) = \lambda_j^{d_j}\partial_{x_j}([\theta']_\nu) = [\partial_{x_j}\theta']_\nu\end{aligned}$$

where $[\phi]_\lambda * [\psi]_\mu = [\theta']_\nu$ by part (a), and now $\partial_{x_j}\theta'$ has cancellation in the variable x_j. Thus the lemma follows in the case of Abelian convolution on $\mathbb{R}^n$.

For convolution $*$ on a homogeneous nilpotent Lie group, in general it is *not* the case that $\partial_{x_j}f * g = f * \partial_{x_j}g$ or that $\partial_{x_j}f * g = \partial_{x_j}[f * g]$. However, we have the following facts.[1]

(a) If X is a left-invariant vector field and Y is a right-invariant vector field then $X[f * g] = f * X[g]$ and $Y[f * g] = Y[f] * g$;

(b) If X is a left-invariant vector field and $Y = \widetilde{X}$ is the unique right-invariant vector field agreeing with X at the origin, then $X[f] * g = f * Y[g]$;

(c) For any $1 \leq j \leq N-1$ we can write

$$\frac{\partial f}{\partial x_j}(\mathbf{x}) = Z_j[f](\mathbf{x}) + \sum_{k=j+1}^{N} P_{j,k}(\mathbf{x})\frac{\partial f}{\partial x_k}(\mathbf{x}) = Z_j[f](\mathbf{x}) + \sum_{k=j+1}^{N}\frac{\partial}{\partial x_k}\big[P_{j,k}(\mathbf{x})f(\mathbf{x})\big]$$

where Z_j is either a left- or a right-invariant vector field, and $P_{j,k}$ is a polynomial of the form

$$P_{j,k}(x_1,\ldots,x_{k-1}) = \sum_{\alpha\in\mathfrak{A}(j,k)} c_{j,k}(\alpha)x_1^{\alpha_1}\cdots x_{k-1}^{\alpha_{k-1}}$$

where $\mathfrak{A}(j,k) = \big\{(\alpha_1,\ldots,\alpha_{k-1}) \in \mathbb{Z}^{k-1} : \alpha_1 d_1 + \cdots + \alpha_{k-1}d_{k-1} = d_k - d_j\big\}$

.

[1]See Proposition 6.16 in [**NRSW12**] or page 22 in [**FS82**].

We can establish part (b) of the lemma by repeating the Abelian argument, but replacing the derivative ∂_{x_j} by the appropriate invariant vector field Z_j. When $j < N$ this introduces an error which is a sum of terms with cancellation in variables x_k with $k > j$. Explicitly,

$$[\partial_{x_j}\phi]_\lambda = \lambda_j^{d_j}\partial_{x_j}\big([\phi]_\lambda\big) = \lambda_j^{d_j} Z_j\big([\phi]_\lambda\big) + \lambda_j^{d_j}\sum_{k=j+1}^{N}\partial_{x_k}\big(P_{j,k}(\mathbf{x})[\phi]_\lambda(\mathbf{x})\big)$$

Now it follows as in Proposition 4.5 of [**NRSW12**] that

$$\begin{aligned}[\partial_{x_j}\phi]_\lambda &= \lambda_j^{d_j} Z_j\big([\phi]_\lambda\big) + \sum_{k=j+1}^{N}\left(\frac{\lambda_j}{\lambda_k}\right)^{d_j}\lambda_k^{d_k}\partial_{x_j}\big([\Theta_{j,k}]_\lambda\big)\\ &= \lambda_j^{d_j} Z_j\big([\phi]_\lambda\big) + \sum_{k=j+1}^{N}\left(\frac{\lambda_j}{\lambda_k}\right)^{d_j}[\partial_{x_j}\Theta_{j,k}]_\lambda\end{aligned}$$

where the functions $\Theta_{j,k}$ are normalized relative to ϕ. The term $\lambda_j^{d_j} Z_j\big([\phi]_\lambda\big)$ is handled as in the Abelian case, since now the vector field Z_j can either be moved across a convolution or put in front of the convolution. This gives a gain of $\left(\frac{\lambda_j}{\mu_j}\right)^{d_j}$. The remaining terms now have cancellation in variables x_k with $k > 1$, and there is a gain $\left(\frac{\lambda_j}{\lambda_k}\right)^{d_j} \leq \left(\frac{\lambda_j}{\lambda_{j+1}}\right)^{d_j}$. Of course when $j = N$, there is no correction term. This completes the sketch. Complete details can be found in Section 6 of [**NRSW12**]. □

9.4. A further decomposition of $\Gamma_{\mathbb{Z}}(\mathbf{E}_S) \times \Gamma_{\mathbb{Z}}(\mathbf{E}_T)$

Fix two marked partitions

$$S = \big\{(I_1,k_1);\ldots;(I_s,k_s)\big\} \qquad \text{and} \qquad T = \{(J_1,\ell_1);\ldots;(J_q,\ell_q)\}$$

in $\mathcal{S}'(\mathbf{E})$. We return to the study of the infinite sum $\sum_{(I,J)\in\Gamma_{\mathbb{Z}}(\mathbf{E}_S)\times\Gamma_{\mathbb{Z}}(\mathbf{E}_T)}[\varphi^I]_I * [\psi^J]_J$.[2] To analyze this, we need a further decomposition of the index set $\Gamma_{\mathbb{Z}}(\mathbf{E}_S)\times\Gamma_{\mathbb{Z}}(\mathbf{E}_T)$. Recall from equation (1.15) that

$$\begin{aligned}[\varphi^I]_I(\mathbf{x}) &= 2^{-\sum_{r=1}^{s} i_r\widehat{Q}_{S,r}}\varphi^I\big(2^{-i_1}\,\hat{\cdot}_S\,\mathbf{x}_{I_1},\ldots,2^{-i_s}\,\hat{\cdot}_S\,\mathbf{x}_{I_s}\big),\\ [\psi^J]_J(\mathbf{x}) &= 2^{-\sum_{a=1}^{q} j_a\widehat{Q}_{T,a}}\psi^J\big(2^{-j_1}\,\hat{\cdot}_T\,\mathbf{x}_{J_1},\ldots,2^{-j_q}\,\hat{\cdot}_T\,\mathbf{x}_{J_q}\big),\end{aligned} \tag{9.3}$$

where, using the identities $e(k_r,j) = \tau_S(j,k_r)^{-1}$ and $e(\ell_a,l) = \tau_T(l,\ell_a)^{-1}$ from Proposition 3.8, together with (3.6) and Definition 3.12 (b), we have

$$\begin{aligned}\widehat{Q}_{S,r} &= \sum_{j\in I_r} Q_j\,e(k_r,j)^{-1} &&= \sum_{j\in I_r} Q_j\tau_S(j,k_r),\\ 2^{-i_r}\,\hat{\cdot}_S\,\mathbf{x}_{I_r} &= \Big\{2^{-i_r/e(k_r,j)}\cdot\mathbf{x}_v : j\in I_r\Big\} &&= \Big\{2^{-i_r\tau_S(j,k_r)}\cdot\mathbf{x}_v : j\in I_r\Big\},\\ \widehat{Q}_{T,a} &= \sum_{l\in J_a} Q_a e(\ell_a,l)^{-1} &&= \sum_{l\in J_a} Q_a\tau_T(l,\ell_a),\\ 2^{-j_a}\,\hat{\cdot}_T\,\mathbf{x}_{J_a} &= \Big\{2^{-j_a/e(\ell_a,v)}\cdot\mathbf{x}_v : v\in J_a\Big\} &&= \Big\{2^{-j_a\tau_T(v,\ell_a)}\cdot\mathbf{x}_v : v\in J_a\Big\}.\end{aligned} \tag{9.4}$$

---

[2]Since S and T are fixed, we now write φ^I and ψ^J instead of φ^I_S and ψ^J_T.

According to Lemma 9.4, for each index j, $1 \le j \le n$, the scale of the support of the convolution $[\varphi^I]_I * [\psi^J]_J$ in the direction of each coordinate $\mathbf{x}_j$ is the maximum of the scales of the supports of each factor. We let $1 \le r \le s$ and $1 \le a \le q$ and focus on indices $j \in I_r \cap J_a$ provided that the intersection is non-empty. The size of the support of $[\varphi^I]_I$ in the direction of $\mathbf{x}_j$ is on the order of $2^{i_r \tau_S(j,k_r)}$, while the size of the support of $[\psi^J]_J$ is on the order of $2^{j\tau_T(v_j,\ell_a)}$. Since the indices i_r and j_a are negative, we have

$$2^{i_r\tau_S(j,k_r)} \ge 2^{j_a\tau_T(j,\ell_a)} \iff i_r\tau_S(j,k_r) \ge j_a\tau_T(j,\ell_a)$$
$$\iff \frac{i_r}{j_a} \le \frac{\tau_T(j,\ell_a)}{\tau_S(j,k_r)} = \frac{e(k_r,j)}{e(\ell_a,l)}.$$

Write[3] $I_r \cap J_a = \{v_1, \dots, v_m\}$, and order these indices so that

$$0 < \frac{\tau_T(v_1,\ell_a)}{\tau_S(v_1,k_r)} \le \frac{\tau_T(v_2,\ell_a)}{\tau_S(v_2,k_r)} \le \cdots \le \frac{\tau_T(v_{m-1},\ell_a)}{\tau_S(v_{m-1},k_r)} \le \frac{\tau_T(v_m,\ell_a)}{\tau_S(v_m,k_r)} < \infty. \tag{9.5}$$

The ratios in equation (9.5) divide the positive real axis into at most $m+1$ subintervals. For any $I = (i_1, \dots, i_s) \in \Gamma_{\mathbb{Z}}(\mathbf{E}_S)$ and $J = (j_1, \dots, j_q) \in \Gamma_{\mathbb{Z}}(\mathbf{E}_T)$, the ratio i_r/j_a must lie in one of these intervals. We partition $\Gamma_{\mathbb{Z}}(\mathbf{E}_S) \times \Gamma_{\mathbb{Z}}(\mathbf{E}_T)$ into disjoint sets so that for (I, J) in each set, the ratios i_r/j_a always lie in the same interval.

To describe and index this decomposition, we proceed as follows. Let

$$\Sigma_{S,T} = \{(I_r, J_a) : 1 \le r \le s,\ 1 \le a \le q,\ I_r \cap J_a \ne \emptyset\},$$

For each $(I_r, J_a) \in \Sigma_{S,T}$ consider the set of $m_{r,a}+1$ intervals

(9.6)
$$\Bigg\{\Big[0, \frac{\tau_T(v_1,\ell_a)}{\tau_S(v_1,k_r)}\Big) = E_0^{r,a}, \Big[\frac{\tau_T(v_1,\ell_a)}{\tau_S(v_1,k_r)}, \frac{\tau_T(v_2,\ell_a)}{\tau_S(v_2,k_r)}\Big) = E_1^{r,a}, \dots$$
$$\dots, \Big[\frac{\tau_T(v_j,\ell_a)}{\tau_S(v_j,k_r)}, \frac{\tau_T(v_{j+1},\ell_a)}{\tau_S(v_{j+1},k_r)}\Big) = E_j^{r,a}, \dots$$
$$\dots, \Big[\frac{\tau_T(v_{m_{r,a}-1},\ell_a)}{\tau_S(v_{m_{r,a}-1},k_r)}, \frac{\tau_T(v_{m_{r,a}},\ell_a)}{\tau_S(v_{m_{r,a}},k_r)}\Big) = E_{m_{r,a}-1}^{r,a}, \Big[\frac{\tau_T(v_{m_{r,a}},\ell_a)}{\tau_S(v_{m_{r,a}},k_r)}, \infty\Big) = E_{m_{r,a}}^{r,a}\Bigg\},$$

and let

$$\widetilde{\mathcal{W}}(S,T) = \prod_{(I_r,J_a)\in\Sigma_{S,T}} \{E_0^{r,a}, E_1^{r,a}, \dots, E_{m_{r,a}}^{r,a}\},$$

i.e., the set of functions which assign to each $(I_r, J_a) \in \Sigma_{S,T}$ one of the intervals in (9.6). Hence, if $W \in \widetilde{\mathcal{W}}(S,T)$ and $(I_r, J_a) \in \Sigma_{S,T}$, $W(I_r, J_a)$ is one of these intervals, and we write

$$W(I_r, J_a) = E_{w(I_r,J_a)}^{r,a}$$

so that $w(I_r, J_a) \in \{0, 1, \dots, m_{r,a}\}$. (Thus $W(I_r, J_a)$ is one of the $m_{r,a}+1$ intervals, while $w(I_r, J_a)$ is the number of that interval.) Setting $j = w(I_r, J_a)$, let

$$\Lambda_W(J_a, I_r) = \begin{cases} \infty & \text{if } j = 0, \\ \frac{\tau_S(v_j,k_r)}{\tau_T(v_j,\ell_a)} & \text{if } j > 0, \end{cases} \qquad \Lambda_W(I_r, J_a) = \begin{cases} \frac{\tau_T(v_{j+1},\ell_a)}{\tau_S(v_{j+1},k_r)} & \text{if } j < m_{r,a}, \\ \infty & \text{if } j = m_{r,a}, \end{cases}$$

[3] We should write $I_r \cap J_a = \{v_1^{r,a}, \dots, v_{m(r,a)}^{r,a}\}$, but as long as the context is clear we omit the dependence on (r, a) for simplicity of notation.

so that $\Lambda_W(J_a, I_r)^{-1}$ and $\Lambda_W(I_r, J_a)$ are the left and right endpoints of $E_{w(I_r,J_a)}$.

We now construct the desired partition. For $W \in \widetilde{\mathcal{W}}(S,T)$, put

(9.7)

$$\Gamma_W(S,T) = \Big\{ \big((i_1,\dots,i_s),(j_1,\dots,j_q)\big) \in \Gamma_{\mathbb{Z}}(\mathbf{E}_S) \times \Gamma_{\mathbb{Z}}(\mathbf{E}_T) : \\ (I_r, J_a) \in \Sigma_{S,T} \Longrightarrow \frac{1}{\Lambda_W(J_a, I_r)} \leq \frac{i_r}{j_a} < \Lambda_W(I_r, J_a) \Big\}.$$

It may happen that for some $W \in \widetilde{\mathcal{W}}(S,T)$ the set $\Gamma_W(S,T)$ is empty[4], and so we put

$$\mathcal{W}(S,T) = \Big\{ W \in \widetilde{\mathcal{W}}(S,T) : \Gamma_W(S,T) \neq \emptyset \Big\}.$$

The following Proposition is an immediate consequence of the definition of $\Gamma_W(S,T)$ and the ordering in equation (9.5).

PROPOSITION 9.5. Let $S = \big\{(I_1,k_1);\dots;(I_s,k_s)\big\}$, $T = \{(J_1,\ell_1);\dots;(J_q,\ell_q)\}$ belong to $\mathcal{S}'(\mathbf{E})$.

(a) The sets $\{\Gamma_W(S,T) : W \in \mathcal{W}(S,T)\}$ partition the set $\Gamma_{\mathbb{Z}}(\mathbf{E}_S) \times \Gamma_{\mathbb{Z}}(\mathbf{E}_T)$:

$$\Gamma_{\mathbb{Z}}(\mathbf{E}_S) \times \Gamma_{\mathbb{Z}}(\mathbf{E}_T) = \bigcup_{W \in \mathcal{W}(S,T)} \Gamma_W(S,T),$$

$$W_1, W_2 \in \mathcal{W}(S,T), \quad W_1 \neq W_2 \Longrightarrow \Gamma_{W_1}(S,T) \cap \Gamma_{W_2}(S,T) = \emptyset.$$

(b) Let $W \in \mathcal{W}(S,T)$ and suppose $I_r \cap J_a = \{v_1,\dots,v_m\} \neq \emptyset$, ordered as in equation (9.5). Suppose that $W(I_r,J_a) = E_{w(I_r,J_a)}$. Let

$$\big((i_1,\dots,i_s),(j_1,\dots,j_q)\big) \in \Gamma_W(S,T).$$

Then for $1 \leq k \leq m$,

$$k \leq w(I_r,J_a) \Longleftrightarrow j_a\tau_T(v_k,\ell_a) \geq i_r\tau_S(v_k,k_r)$$
$$k > w(I_r,J_a) \Longleftrightarrow j_a\tau_T(v_k,\ell_a) < i_r\tau_S(v_k,k_r)$$

In particular,

(1) if $w(I_r,J_a) = 0$, then $i_r\tau_S(v_k,k_r) > j_a\tau_T(v_k,\ell_a)$ for all $1 \leq k \leq m$;

(2) if $w(I_r,J_a) = m$ then $i_r\tau_S(v_k,k_r) \leq j_a\tau_T(v_k,\ell_a)$ for all $1 \leq k \leq m$;

(3) if $k_r \in I_r \cap J_a$ then

$$k_r \leq w(I_r,J_a) \Longrightarrow i_r \leq \tau_T(k_r,\ell_a)\, j_a < 0,$$
$$k_r > w(I_r,J_a) \Longrightarrow \tau_T(k_r,\ell_a)\, j_a < i_r < 0;$$

(4) if $\ell_a \in I_r \cap J_a$ then

$$\ell_a > w(I_r,J_a) \Longrightarrow j_a < \tau_S(\ell_a,k_r)\, i_r < 0,$$
$$\ell_a \leq w(I_r,J_a) \Longrightarrow \tau_S(\ell_a,k_r)\, i_r \leq j_a < 0.$$

[4]For instance, if there are equalities among the weak inequalities in (9.5), some of the intervals are empty.

To summarize, we have partitioned $\Gamma_{\mathbb{Z}}(\mathbf{E}_S) \times \Gamma_{\mathbb{Z}}(\mathbf{E}_T)$ into a disjoint union of sets $\Gamma_W(S,T)$. Let $I = (i_1, \ldots, i_s)$ and $J = (j_1, \ldots, j_q)$. Then $(I,J) \in \Gamma_W(S,T)$ if and only if

$$
\begin{aligned}
e_S(p,r)i_r \le i_p &< 0 && \text{for } r,p \in \{1,\ldots,s\},\\
e_T(b,a)j_a \le j_b &< 0 && \text{for } a,b \in \{1,\ldots,q\},\\
\Lambda_W(I_r,J_a)j_a &< i_r && \text{if } I_r \cap J_a \neq \emptyset,\\
\Lambda_W(J_a,I_r)i_r &\le j_a && \text{if } I_r \cap J_a \neq \emptyset.
\end{aligned}
\tag{9.8}
$$

It can happen that $\Lambda_W(I_r,J_a)$ or $\Lambda_W(J_a,I_r)$ is $+\infty$, in which case there is simply no corresponding inequality between i_r and j_a. However, if there is no bound of an index i_r by *any* j_a, then $\Lambda_W(I_r,J_a) = \infty$ for every a such that $I_r \cap J_a \neq \emptyset$, and this means that the scale of the dilation by i_r is *always* smaller than the scale of the dilation by the corresponding j_a. Similarly, if there is no bound of j_a by *any* i_r, then the scale of the dilation by j_a is *always* less than or equal to the scale of of the dilation by the corresponding i_r.[5]

9.5. Fixed and free indices

Continue to fix $S = \big\{(I_1,k_1);\ldots;(I_s,k_s)\big\}$ and $T = \{(J_1,\ell_1);\ldots,(J_q,\ell_q)\}$ in $\mathcal{S}'(\mathbf{E})$. The index set $\mathcal{W}(S,T)$ is finite, so formally, using the partition from Proposition 9.5, we have

$$
\begin{aligned}
\mathcal{K}_S * \mathcal{L}_T(\mathbf{x}) &= \sum_{W\in\mathcal{W}(S,T)} \left[\sum_{(I,J)\in\Gamma_W(S,T)} [\varphi^I]_I * [\psi^J]_J(\mathbf{x})\right]\\
&= \sum_{W\in\mathcal{W}(S,T)} \mathcal{R}_{S,T,W}(\mathbf{x}).
\end{aligned}
$$

Thus to prove Lemma 9.3 (and hence prove Theorem 9.2) it suffices to show that for each $W \in \mathcal{W}(S,T)$ the inner infinite sum $\mathcal{R}_{S,T,W} = \sum_{(I,J)\in\Gamma_W(S,T)} [\varphi^I]_I * [\psi^J]_J$ converges in the sense of distributions, and that the sum belongs to the class $\mathcal{P}_0(\mathbf{E})$. We will do this by showing that there is a matrix $\mathbf{E}(S,T,W)$ so that $\mathcal{R}_{S,T,W} \in \mathcal{P}_0(\mathbf{E}(S,T,W)) \subset \mathcal{P}_0(\mathbf{E})$.

DEFINITION 9.6. Let $S = \big\{(I_1,k_1);\ldots;(I_s,k_s)\big\}, T = \{(J_1,\ell_1);\ldots,(J_q,\ell_q)\} \in \mathcal{S}'(\mathbf{E})$ and $W \in \mathcal{W}(S,T)$. Recall that if $(I_r,J_a) \in \Sigma_{S,T}$ then $I_r \cap J_a \neq \emptyset$ and $W(I_r,J_a) = E_{w(I_r,J_a)}$ where $\{E_0,\ldots,E_m\}$ are the intervals listed in equation (9.6).

[5] In the notation of the next section, this means that in the first case, i_r is a free index, and in the second case j_a is a free index. See Definition 9.6.

(a) Set

$$
\begin{aligned}
I_r^* &= \bigcup_{\substack{a=1\\ I_r\cap J_a\neq\emptyset}}^{q} \Big\{ j\in I_r\cap J_a : i_r\,\tau_S(j,k_r) > j_a\,\tau_T(j,\ell_a) \\
&\qquad\qquad \forall\big((i_1,\dots,i_s),(j_1,\dots,j_q)\big)\in\Gamma_W(S,T)\Big\}\\
&= \bigcup_{\substack{a=1\\ I_r\cap J_a\neq\emptyset}}^{q} \Big\{ v_k\in I_r\cap J_a : k> w(I_r,J_a)\Big\};\\
J_a^* &= \bigcup_{\substack{r=1\\ I_r\cap J_a\neq\emptyset}}^{w} \Big\{ j\in I_r\cap J_a : i_r\,\tau_S(j,k_r) \le j_a\,\tau_T(j,\ell_a) \\
&\qquad\qquad \forall\big((i_1,\dots,i_s),(j_1,\dots,j_q)\big)\in\Gamma_W(S,T)\Big\}\\
&= \bigcup_{\substack{a=1\\ I_r\cap J_a\neq\emptyset}}^{q} \Big\{ v_k\in I_r\cap J_a : k\le w(I_r,J_a)\Big\}.
\end{aligned}
\tag{9.9}
$$

(b) An index r, $1\le r\le s$ is *fixed* if $I_r^*\neq\emptyset$, and is *free* if $I_r^*=\emptyset$.

(c) An index a, $1\le a\le q$ is *fixed* if $J_a^*\neq\emptyset$, and is *free* if $J_a^*=\emptyset$.

REMARK 9.7. The notions of fixed and free indices and the definitions of the sets I_r^* and J_a^* depend on the choice of $S,T\in\mathcal{S}'(\mathbf{E})$ and on the choice of $W\in\mathcal{W}(S,T)$. In particular, they are the same for all choices of $(I,J)\in\Gamma_W(S,T)$.

Clearly $I_r^*\subset I_r$, $J_a^*\subset J_a$ and $I_r^*\cap J_a^*=\emptyset$. Also, for every $j\in\{1,\dots,n\}$ there is a unique pair $(I_r,J_a)\in\Sigma_{S,T}$ so that $j\in I_r\cap J_a$, and then either $j\in I_r^*$ or $j\in J_a^*$. Thus we have a disjoint decomposition

$$
\{1,\dots,n\}=\bigcup_{r=1}^{s} I_r^*\cup\bigcup_{a=1}^{q} J_a^*. \tag{9.10}
$$

The following Proposition provides an explanation of the significance of fixed and free indices. The proof just involves unwinding the definitions.

PROPOSITION 9.8.

(a) An index $r\in\{1,\dots,s\}$ is free if and only if for every $j\in I_r$ and every $(I,J)\in\Gamma_W(S,T)$, the scale of the support of $[\varphi^I]_I$ is smaller in the direction of $\mathbf{x}_j$ than the scale of the support of $[\psi^J]_J$.

(b) An index $r\in\{1,\dots,s\}$ is fixed if and only if there exists at least one $j\in I_r$ so that for every $(I,J)\in\Gamma_W(S,T)$, the scale of the support of $[\varphi^I]_I$ in the direction of $\mathbf{x}_j$ is larger than the scale of the support of $[\psi^J]_J$.

(c) The indices $j\in\{1,\dots,n\}$ for which the scale of the support of $[\varphi^I]_I$ is larger than the scale of the support of $[\psi^J]_J$ in the direction of $\mathbf{x}_j$ is precisely $\bigcup_{r=1}^{s} I_r^*$.

(d) An index $a\in\{1,\dots,q\}$ is free if and only if for every $j\in J_a$ and for every $(I,J)\in\Gamma_W(S,T)$, the scale of the support of $[\psi^J]_J$ in the direction of $\mathbf{x}_j$ is smaller than the scale of the support of $[\varphi^I]_I$.

(e) An index $a \in \{1, \ldots, q\}$ is fixed if and only if there exists at least one $j \in J_a$ so that for every $(I, J) \in \Gamma_W(S, T)$ the scale of the support of $[\psi^J]_J$ is larger than the scale of the support of $[\varphi^I]_I$ in the direction of $\mathbf{x}_j$.

(f) The indices $j \in \{1, \ldots, n\}$ for which the scale of the support of $[\psi^J]_J$ is larger than the scale of the support of $[\varphi^I]_I$ in the direction of $\mathbf{x}_j$ is precisely $\bigcup_{a=1}^q J_a^*$.

Let $S^* = \{r_1, \ldots, r_\alpha\}$ denote the subset of $\{1, \ldots, s\}$ consisting of fixed indices, and let $T^* = \{a_1, \ldots, a_\beta\}$ denote the subset of $\{1, \ldots, q\}$ consisting of fixed indices. It may happen that $\alpha = 0$ or that $\beta = 0$, but we always have $\alpha + \beta \geq 1$. Then we can write

$$\{1, \ldots, n\} = \bigcup_{r \in S^*} I_r^* \cup \bigcup_{a \in T^*} J_a^* = \bigcup_{j=1}^{\alpha} I_{r_j}^* \cup \bigcup_{k=1}^{\beta} J_{a_k}^*. \tag{9.11}$$

9.6. A finer decomposition of $\mathbb{R}^N$

Using the decomposition in equation (9.11) we study the new decomposition of $\mathbb{R}^N$ indexed by the fixed indices in $\{1, \ldots, s\}$ and $\{1, \ldots, q\}$:

$$\mathbb{R}^N = \bigoplus_{r \in S^*} \mathbb{R}^{I_r^*} \oplus \bigoplus_{a \in T^*} \mathbb{R}^{J_a^*} = \bigoplus_{j=1}^{\alpha} \mathbb{R}^{I_{r_j}^*} \oplus \bigoplus_{j=1}^{\beta} \mathbb{R}^{J_{a_j}^*}$$

where

$$\mathbb{R}^{I_r^*} = \big\{\mathbf{x} = (\mathbf{x}_1, \ldots, \mathbf{x}_n) \in \mathbb{R}^N : \mathbf{x}_j \neq 0 \Longrightarrow j \in I_r^*\big\},$$
$$\mathbb{R}^{J_a^*} = \big\{\mathbf{x} = (\mathbf{x}_1, \ldots, \mathbf{x}_n) \in \mathbb{R}^N : \mathbf{x}_j \neq 0 \Longrightarrow j \in J_a^*\big\}.$$

Write elements of $\mathbb{R}^{I_r^*}$ and $\mathbb{R}^{J_a^*}$ as $\mathbf{x}_{I_r^*} = \{\mathbf{x}_j : j \in I_r^*\}$ and $\mathbf{x}_{J_a^*} = \{\mathbf{x}_j : j \in J_a^*\}$. This gives us a set of coordinates on $\mathbb{R}^N$.

Order these coordinates so that the corresponding set of indices

$$\{k_{r_1}, \ldots, k_{r_\alpha}, \ell_{a_1}, \ldots, \ell_{a_\beta}\}$$

together are increasing. With this ordering, rename the subspaces as

$$\{\mathbb{R}^{V_1}, \ldots, \mathbb{R}^{V_{\alpha+\beta}}\}.$$

Then every element $\mathbf{x} \in \mathbb{R}^N$ is an $(\alpha+\beta)$-tuple $\mathbf{x} = (\overline{\mathbf{x}}_1, \ldots, \overline{\mathbf{x}}_m, \ldots, \overline{\mathbf{x}}_{\alpha+\beta})$ where $\overline{\mathbf{x}}_m \in \mathbb{R}^{V_k}$ and V_k is either I_r^* for some $r \in S^*$ or J_a^* for some $a \in T^*$. By the choice of the ordering, if $m_1 < m_2$ then

$$\begin{aligned}
V_{m_1} &= I_r^*, & V_{m_2} &= I_p^* & &\Longrightarrow & k_r &< k_p,\\
V_{m_1} &= I_r^*, & V_{m_2} &= J_a^* & &\Longrightarrow & k_r &< \ell_a,\\
V_{m_1} &= J_a^*, & V_{m_2} &= I_r^* & &\Longrightarrow & \ell_a &< k_r,\\
V_{m_1} &= J_a^*, & V_{m_2} &= J_b^* & &\Longrightarrow & \ell_a &< \ell_b.
\end{aligned}$$

On each component $\mathbb{R}^{V_m}$ we use the dilation structure inherited from either I_r or J_a. Thus from (9.4) we define a dilation which we write $\circ$ by setting

$$\lambda \circ \overline{\mathbf{x}}_m = \begin{cases} \Big\{\lambda^{\tau_S(j,k_r)} \cdot \mathbf{x}_j : j \in I_r^*\Big\} & \text{if } \overline{\mathbf{x}}_m = \mathbf{x}_{I_r^*} \\[2ex] \Big\{\lambda^{\tau_T(j,\ell_a)} \cdot \mathbf{x}_j : j \in J_a^*\Big\} & \text{if } \overline{\mathbf{x}}_m = \mathbf{x}_{J_a^*} \end{cases}. \tag{9.12}$$

(To keep notation simple we write $\circ$ instead of $\circ_{S,T,W}$.) The corresponding homogeneous norms are given by

$$n_{S,r}(\mathbf{x}_{I_r^*}) = \sum_{j\in I_r^*} n_j(\mathbf{x}_j)^{1/\tau_S(j,k_r)}, \qquad n_{T,a}(\mathbf{x}_{J_a^*}) = \sum_{j\in J_a^*} n_j(\mathbf{x}_j)^{1/\tau_T(j,\ell_a)},$$

and the homogeneous dimension of $\mathbb{R}^{V_m}$ is then

$$Q_m = \begin{cases} Q_{I_r^*} = \sum_{j\in I_r^*} Q_j\, \tau_S(j,k_r) & \text{if } \overline{\mathbf{x}}_m = \mathbf{x}_{I_r^*}; \\ \\ Q_{J_a^*} = \sum_{j\in J_a^*} Q_j\, \tau_T(j,\ell_a) & \text{if } \overline{\mathbf{x}}_m = \mathbf{x}_{J_a^*}. \end{cases}$$

9.7. The matrix $\mathbf{E}_{S,T,W}$

If $\big((i_1,\dots,i_s),(j_1,\dots,j_q)\big) \in \Gamma_W(S,T) \subset \Gamma_{\mathbb{Z}}(\mathbf{E}_S)\times\Gamma_{\mathbb{Z}}(\mathbf{E}_T)$, we obtain an $(\alpha+\beta)$-tuple $\{i_{r_1},\dots,i_{r_\alpha},j_{a_1},\dots,j_{a_\beta}\}$ by keeping only the i_r and j_s corresponding to fixed indices S^* and T^*. Let $\pi:\mathbb{Z}^{s+q}\to\mathbb{Z}^{\alpha+\beta}$ be the mapping which takes $(i_1,\dots,i_s),(j_1,\dots,j_q)\in\mathbb{Z}^{s+q}$ to $(i_{r_1},\dots,i_{r_\alpha},j_{a_1},\dots,j_{a_\beta})\in\mathbb{Z}^{\alpha+\beta}$. Our next objective is to describe the inequalities satisfied by the entries of these $(\alpha+\beta)$-tuples.

It follows from equation (9.8) that

$$\begin{aligned} e_S(r_j,r_k)i_{r_k} &\le i_{r_j} < 0 && \text{for } j,k\in\{1,\dots,\alpha\},\\ e_T(a_l,a_m)j_{a_m} &\le j_{a_l} < 0 && \text{for } l,m\in\{1,\dots,\beta\},\\ \Lambda_W(I_{r_j},J_{a_m})j_{a_m} &< i_{r_j} && \text{if } I_{r_j}\cap J_{a_m}\neq\emptyset,\\ \Lambda_W(J_{a_m},I_{r_j})i_{r_j} &\le j_{a_m} && \text{if } I_{r_j}\cap J_{a_m}\neq\emptyset. \end{aligned} \tag{9.13}$$

Let

$$\Gamma(S,T,W) = \Big\{(i_{r_1},\dots,i_{r_\alpha},j_{a_1},\dots,j_{a_\beta})\in\mathbb{Z}^{\alpha+\beta} \text{ which satisfy (9.13)}\Big\}. \tag{9.14}$$

Thus $\pi:\Gamma_W(S,T)\to\Gamma(S,T,W)$.

PROPOSITION 9.9. The mapping $\pi:\Gamma_W(S,T)\to\Gamma(S,T,W)$ is onto.

PROOF. Thus suppose that the $(\alpha+\beta)$-tuple $(i_{r_1},\dots,i_{r_\alpha},j_{a_1},\dots,j_{a_\beta})$ satisfies the inequalties in equation (9.13). First observe that the first two lines of inequalities describe the projections of $\Gamma(\mathbf{E}_S)$ and $\Gamma(\mathbf{E}_T)$ onto the smaller sets of coordinates. Since $\mathbf{E}_S$ and $\mathbf{E}_T$ satisfy the basic hypotheses (2.10), it follows from Lemma 14.4 that there exists $(i_1,\dots,i_s)\in\Gamma(\mathbf{E}_S)$ and $(t_1,\dots,t_q)\in\Gamma(\mathbf{E}_T)$ such that

$$\pi(i_1,\dots,i_s,j_1,\dots,j_q) = (i_{r_1},\dots,i_{r_\alpha},j_{a_1},\dots,j_{a_\beta}).$$

We claim that $(i_1,\dots,i_s,j_1,\dots,j_q)\in\Gamma_W(S,T)$, so we need to check all the inequalities in equation (9.8). The first two lines are clearly satisfied, since they only depend on the fact that $(i_1,\dots,i_s)\in\Gamma(\mathbf{E}_S)$ and $(j_1,\dots,j_q)\in\Gamma(\mathbf{E}_T)$. Also the second two lines are satisfied provided that $r=r_j$ and $a=a_m$ for some $j\in\{1,\dots,\alpha\}$ and $m\in\{1,\dots,\beta\}$. It remains to check what happens if i_r is a free index or j_a is a free index. But as we observed in the summary on page 79, if i_r is a free index and $I_r\cap J_a\neq\emptyset$ then $\Lambda_W(I_r,J_a)=\infty$ so there is no third line in the inequalities. Also if i_r is a free index, then the scale of the dilation $\tau_S(j,k_r)i_r$ must be smaller than the scale of the dilation $\tau_T(j,\ell_a)j_a$ for every $j\in I_r\cap J_a$. This says that the ratio i_r/j_a must lie in the largest (*i.e.* the unbounded) interval, and this is the statement of the fourth inequality. A similar argument works if j_a is a free index. $\square$

The cone $\Gamma(S,T,W) \subset \mathbb{R}^{\alpha+\beta}$ is defined by the inequalities in (9.13), which, using the notation in Section 14.2, is defined by a *partial matrix* $\mathbf{B}_W$, with indices varying in the disjoint union $S^* \sqcup T^*$ and given by

$$b(\mu,\nu) = \begin{cases} e_S(\mu,\nu) & \text{if } \mu,\nu \in S^* \\ e_T(\mu,\nu) & \text{if } \mu,\nu \in T^* \\ \Lambda_W(I_\mu, J_\nu) & \text{if } \mu \in S^*,\ \nu \in T^* \\ \Lambda_W(J_\mu, I_\nu) & \text{if } \mu \in T^*,\ \nu \in S^*. \end{cases} \tag{9.15}$$

LEMMA 9.10. *The matrix* $\mathbf{B}_W$ *is connected.*

PROOF. We must prove that, given $\mu \neq \nu \in S^* \sqcup T^*$, there are $\lambda_1, \dots, \lambda_a \in S^* \sqcup T^*$ such that

$$b_W(\mu,\lambda_1) b_W(\lambda_1,\lambda_2) \cdots b_W(\lambda_a,\nu) < \infty.$$

There is no doubt that this holds if both μ and ν belong either to S^* or to T^* because all the e_S and e_T are finite. For the other two cases it is necessary and sufficient to prove that there exist $r \in S^*$, $a \in T^*$ with $\Lambda_W(I_r, J_a) < \infty$ and $r' \in S^*$, $a' \in T^*$ with $\Lambda_W(J_{a'}, I_{r'}) < \infty$ (equivalently with $w(I_r, J_a) \neq 0$ and $w(I_{r'}, J_{a'}) \neq m_{r',a'}$).

Assume that one of these conditions is violated, e.g., that $w(I_r, J_a) = 0$ for all $r \in S^*, a \in T^*$. By (9.9), this implies that $I_r \cap J_a \subset I_r^*$ for every r, a. Hence $T^* = \emptyset$ and the proof is completed. □

By Lemma 14.3, there is matrix $\mathbf{E}_{S,T,W}$ which satisfies the basic hypotheses (2.10) so that we have $\Gamma(S,T,W) = \Gamma(\mathbf{E}_{S,T,W})$.

PROPOSITION 9.11. If $(m_1, \dots, m_{\alpha+\beta}) \in \Gamma_{\mathbb{Z}}(\mathbf{E}_{S,T,W})$, the corresponding dilation of $\mathbb{R}^N$ given by

$$(\overline{\mathbf{x}}_1, \dots, \overline{\mathbf{x}}_{\alpha+\beta}) \longrightarrow (2^{-m_1} \circ \overline{\mathbf{x}}_1, \dots, 2^{-m_{\alpha+\beta}} \circ \overline{\mathbf{x}}_{\alpha+\beta})$$

is still compatible (in the sense of Section 8.1) with the nilpotent Lie group structure.

PROOF. Let $1 \leq l_1 < l_2 \leq n$. Then $\mathbf{x}_{l_1}$ and $\mathbf{x}_{l_2}$ are coordinates that belong to one of the sets $\{I_{r_j}^*, 1 \leq j \leq \alpha\}$ or $\{J_{a_k}^*, 1 \leq k \leq \beta\}$. Write the $(\alpha+\beta)$-tuple $(m_1, \dots, m_{\alpha+\beta}) = (i_{r_1}, \dots, i_{r_\alpha}, j_{a_1}, \dots, j_{a_q})$. There are thus four cases to consider.

If $\mathbf{x}_{l_1} \in I_{r_{l_1}}^*$ and $\mathbf{x}_{l_2} \in I_{r_{l_2}}^*$, we proceed as in Lemma 8.6. After taking logarithm to base 2 the scales of the corresponding dilations are $i_{r_{l_1}} \tau_S(l_1, k_{r_{l_1}})$ and $i_{r_{l_2}} \tau_S(l_2, k_{r_{l_2}})$. Now $e_S(r_{l_2}, r_{l_1}) i_{r_{l_1}} \leq i_{r_{l_2}} < 0$ so

$$\begin{aligned} \frac{i_{r_{l_2}}}{i_{r_{l_1}}} &\leq e_S(r_{l_2}, r_{l_1}) \leq \tau_S(k_{r_{l_2}}, k_{r_{l_1}}) = \min_{j \in I_{r_{l_1}}} \frac{e(k_{r_{l_2}}, j)}{e(k_{r_{l_1}}, j)} \\ &\leq \frac{e(k_{r_{l_2}}, l_1)}{e(k_{r_{l_1}}, l_1)} \leq \frac{e(k_{r_{l_2}}, l_2) e(l_2, l_1)}{e(k_{r_{l_1}}, l_1)} \leq \frac{e(k_{r_{l_2}}, l_2)}{e(k_{r_{l_1}}, l_1)} \end{aligned}$$

since $e(l_2, l_1) \leq 1$. Thus

$$i_{r_{l_1}} \tau_S(l_1, k_{r_{l_1}}) = \frac{i_{r_{l_1}}}{e(k_{r_{l_1}}, l_1)} \leq \frac{i_{r_{l_2}}}{e(k_{r_{l_2}}, l_2)} = i_{r_{l_2}} \tau_S(l_2, k_{r_{l_2}}),$$

which is the desired inequality. A similar argument works if $\mathbf{x}_{l_1} \in J_{a_{l_1}}^*$ and $\mathbf{x}_{l_2} \in J_{a_{l_2}}^*$.

Now suppose $\mathbf{x}_{l_1} \in I^*_{r_{l_1}} \cap J_{a_{l_1}}$ and $\mathbf{x}_{l_2} \in I_{r_{l_2}} \cap J^*_{a_{l_2}}$. This time, after taking logarithm to base 2 the scales of the corresponding dilations are $i_{r_{l_1}}\tau_S(l_1, k_{r_{l_1}})$ and $j_{a_{l_2}}\tau_T(l_2, j_{a_{l_2}})$. Just as before, it follows that

$$i_{r_{l_1}}\tau_S(l_1, k_{r_{l_1}}) \le i_{r_{l_2}}\tau_S(l_2, k_{r_{l_2}}).$$

But now since $l_2 \in J^*_{a_{l_2}}$, we know that the i-scale at this variable is less than or equal to the j-scale at this variable. Thus

$$i_{r_{l_2}}\tau_S(l_2, k_{r_{l_2}}) \le j_{a_{l_2}}\tau_T(l_2, l_{a_{l_2}}).$$

Putting the last two inequalities together gives the required estimate. A similar argument works if $\mathbf{x}_{l_1} \in I_{r_{l_1}} \cap J^*_{a_{l_1}}$ and $\mathbf{x}_{l_2} \in I^*_{r_{l_2}} \cap J_{a_{l_2}}$, and this completes the proof. □

9.8. The class $\mathcal{P}_0(\mathbf{E}_{S,T,W})$

Given the decomposition $\mathbb{R}^N = \bigoplus_{j=1}^{\alpha} \mathbb{R}^{I^*_{r_j}} \oplus \bigoplus_{j=1}^{\beta} \mathbb{R}^{j^*_{a_j}}$ and the matrix $\mathbf{E}_{S,T,W}$, we have the space of distributions $\mathcal{P}_0(\mathbf{E}_{S,T,W})$ and the corresponding space of multipliers $\mathcal{M}_\infty(\mathbf{E}_{S,T,W})$. For notational convenience we use the symbol e_W to denote the entries of $\mathbf{E}_{S,T,W}$. Recall that their indices vary in the disjoint union $S^* \sqcup T^*$ of the sets $S^* \subset \{1, \dots, s\}$ and $T^* \subset \{1, \dots, q\}$ of fixed indices relative to W.

LEMMA 9.12. *$\mathcal{M}_\infty(\mathbf{E}_{S,T,W}) \subset \mathcal{M}_\infty(\mathbf{E})$ and consequently $\mathcal{P}_0(\mathbf{E}_{S,T,W}) \subset \mathcal{P}_0(\mathbf{E})$.*

PROOF. The dual norms for the space $\mathcal{M}_\infty(\mathbf{E}_{S,T,W})$ are given by

$$\widehat{N}_{r,W}(\boldsymbol{\xi}) = \sum_{p\in S^*} \widehat{n}_{S,p}(\boldsymbol{\xi}_{I^*_p})^{1/e_W(p,r)} + \sum_{b\in T^*} \widehat{n}_{T,b}(\boldsymbol{\xi}_{J^*_b})^{1/e_W(b,r)}, \qquad r \in S^*;$$
$$\widehat{N}_{a,W}(\boldsymbol{\xi}) = \sum_{p\in S^*} \widehat{n}_{S,p}(\boldsymbol{\xi}_{I^*_p})^{1/e_W(I_p,J_a)} + \sum_{b\in T^*} \widehat{n}_{T,b}(\boldsymbol{\xi}_{J^*_b})^{1/e_W(b,a)}, \qquad a \in T^*.$$

By Corollary 5.4, it suffices to show, for each (j,k), the entries $e^\sharp_W(j,k)$ of the $n \times n$ matrix $\mathbf{E}^\sharp_{S,T,W}$ satisfy the inequality

$$e^\sharp_W(j,k) \le e(j,k).$$

By (5.5), $e^\sharp_W(j,k) = \frac{\alpha_k}{\alpha_j} e_W(\mu,\nu)$, where

(i) $\mu,\nu \in S^* \sqcup T^*$ are such that $j \in I^*_\mu$ (if $\mu \in S^*$) or $j \in J^*_\mu$ (if $\mu \in T^*$) and $k \in I^*_\nu$ or J^*_ν accordingly;

(ii) α_k is the exponent of the norm of $\boldsymbol{\xi}_k$ in

$$\widehat{n}_{S,p}(\boldsymbol{\xi}_{I^*_p}) = \sum_{k\in I^*_p} n_k(\boldsymbol{\xi}_k)^{e(k_p,k)} \text{ (if } k \in I^*_p),$$
$$\widehat{n}_{T,b}(\boldsymbol{\xi}_{J^*_b}) = \sum_{k\in J^*_b} n_k(\boldsymbol{\xi}_k)^{e(\ell_b,k)} \text{ (if } k \in J^*_b).$$

We then have the inequalities

$$e^\sharp_W(j,k) = \begin{cases} \dfrac{e(k_\nu,k)}{e(k_\mu,j)} e_W(\mu,\nu) & \text{if } \mu,\nu \in S^*\,,\ j \in I^*_\mu\,,\ k \in I^*_\nu \\ \dfrac{e(\ell_\nu,k)}{e(\ell_\mu,j)} e_W(\mu,\nu) & \text{if } \mu,\nu \in T^*\,,\ j \in J^*_\mu\,,\ k \in J^*_\nu \\ \dfrac{e(\ell_\nu,k)}{e(k_\mu,j)} e_W(\mu,\nu) & \text{if } \mu \in S^*\,,\ \nu \in T^*\,,\ j \in I^*_\mu\,,\ k \in J^*_\nu \\ \dfrac{e(k_\nu,k)}{e(\ell_\mu,j)} e_W(\mu,\nu) & \text{if } \mu \in T^*\,,\ \nu \in S^*\,,\ j \in J^*_\mu\,,\ k \in I^*_\nu. \end{cases}$$

If $\mu,\nu \in S^*$ and $j \in I^*_\mu$, $k \in I^*_\nu$, using the inequality $e_W(\mu,\nu) \le b_W(\mu,\nu)$, we have

$$e^\sharp_W(j,k) \le \frac{e(k_\nu,k)}{e(k_\mu,j)} b_W(\mu,\nu) = \frac{e(k_\nu,k)}{e(k_\mu,j)} \min_{l\in I_\nu} \frac{e(k_\mu,l)}{e(k_\nu,l)} \le \frac{e(k_\mu,k)}{e(k_\mu,j)} \le e(j,k),$$

and similarly if $\mu,\nu \in T^*$.

Assume now that $\mu \in S^*$, $\nu \in T^*$, $j \in I^*_\mu$, $k \in J^*_\nu$. In this case the inequality $e_W(\mu,\nu) \le b_W(\mu,\nu)$ may not be sufficient (e.g., we may have $b_W(\mu,\nu) = \infty$). However, we can use the basic inequality

$$e_W(\mu,\nu) \le e_W(\mu,\nu')e_W(\nu',\nu) \le b_W(\mu,\nu')b_W(\nu',\nu) \le \Lambda_W(I_\mu, J_{\nu'})\tau_T(\ell_\nu,\ell_{\nu'}),$$

which holds for all $\nu' \in T^*$. Choosing ν' so that $j \in J_{\nu'}$, we have, with $m = w(\mu,\nu')$,

$$\begin{aligned} e^\sharp_W(j,k) &\le \frac{e(\ell_\nu,k)}{e(k_\mu,j)} \Lambda_W(I_\mu, J_{\nu'})\tau_T(\ell_\nu,\ell_{\nu'}) \\ &= \frac{e(\ell_\nu,k)}{e(k_\mu,j)} \frac{e(k_\mu,v_m)}{e(\ell_{\nu'},v_m)} \min_{i\in J_\nu} \frac{e(\ell_{\nu'},i)}{e(\ell_\nu,i)} \\ &\le \frac{e(\ell_\nu,k)}{e(k_\mu,j)} \frac{e(k_\mu,v_m)}{e(\ell_{\nu'},v_m)} \frac{e(\ell_{\nu'},k)}{e(\ell_\nu,k)} \\ &= \frac{e(k_\mu,v_m)}{e(\ell_{\nu'},v_m)} \frac{e(\ell_{\nu'},j)}{e(k_\mu,j)} \frac{e(\ell_{\nu'},k)}{e(\ell_{\nu'},j)}. \end{aligned} \tag{9.16}$$

Since $j \in I^*_\mu \cap J_{\nu'}$, it follows from (9.9) that $k = v_{m'}$ with $m' > m$ in the ordering (9.5) on $I_\mu \cap J^{\nu'}$, i.e.,

$$\frac{e(k_\mu,v_m)}{e(\ell_{\nu.},v_m)} = \frac{\tau_T(v_m,\ell_{\nu'})}{\tau_S(v_m,k_\mu)} \le \frac{\tau_T(j,\ell_{\nu'})}{\tau_S(j,k_\mu)} = \frac{e(k_\mu,j)}{c(\ell_{\nu'},j)}.$$

Hence

$$e^\sharp_W(j,k) \le \frac{e(\ell_{\nu'},k)}{e(\ell_{\nu'},j)} \le e(j,k).$$

The last case, with $\mu \in T^*$, $\nu \in S^*$, $j \in J^*_\mu$, $k \in I^*_\nu$, is treated similarly. □

9.9. Estimates for $[\varphi]_I * [\psi]_J$

Fix marked partitions $S = \big((I_1,k_1);\ldots;(I_s,k_s)\big)$, $T = \big((J_1,\ell_1);\ldots;(J_q,\ell_q)\big)$, and let $W \in \mathcal{W}(S,T)$. Recall that

$$I_r^* = \bigcup_{\substack{a=1\\ I_r\cap J_a\neq\emptyset}}^{q} \Big\{j \in I_r\cap J_a : \frac{i_r}{e(k_r,j)} > \frac{j_a}{e(\ell_a,j)}\ \forall\big((i_1,\ldots,i_s),(j_1,\ldots,j_q)\big)\in \Gamma_{\mathbf{E}_{S,T,W}}\Big\},$$

$$J_a^* = \bigcup_{\substack{r=1\\ I_r\cap J_a\neq\emptyset}}^{s} \Big\{j \in I_r\cap J_a : \frac{i_r}{e(k_r,j)} \leq \frac{j_a}{e(\ell_a,j)}\ \forall\big((i_1,\ldots,i_s),(j_1,\ldots,j_q)\big)\in \Gamma_{\mathbf{E}_{S,T,W}}\Big\}.$$

Then

$$\begin{aligned} S^* &= \big\{r\in\{1,\ldots,s\} : I_r^*\neq\emptyset\big\} = \{r_1,\ldots,r_\alpha\}\\ T^* &= \big\{a\in\{1,\ldots,q\} : J_a^*\neq\emptyset\big\} = \{a_1,\ldots,a_\beta\}\end{aligned}$$

are the fixed indices. Write $\mathbb{R}^N = \bigoplus_{j=1}^{\alpha+\beta} V_j$ as in section 9.6. Thus if $\mathbf{x}\in\mathbb{R}^N$, $\mathbf{x} = (\overline{\mathbf{x}}_1,\ldots,\overline{\mathbf{x}}_{\alpha+\beta})$ where each $\overline{\mathbf{x}}_m$ is either equal to $\mathbf{x}_{I_r^*} = \big\{\mathbf{x}_j : j\in I_r^*\big\}$ for some $r\in S^*$ or is equal to $\mathbf{x}_{J_a^*} = \big\{\mathbf{x}_j : j\in J_a^*\big\}$ for some $a\in T^*$. In equation (9.12) we introduced a family of dilations on the space V_m by setting

$$\lambda\circ\overline{\mathbf{x}}_m = \begin{cases}\Big\{\lambda^{1/e(k_r,j)}\cdot\mathbf{x}_j : j\in I_r^*\Big\} & \text{if } \overline{\mathbf{x}}_m = \mathbf{x}_{I_r^*},\\[2ex] \Big\{\lambda^{1/e(\ell_a,j)}\cdot\mathbf{x}_j : j\in J_a^*\Big\} & \text{if } \overline{\mathbf{x}}_m = \mathbf{x}_{J_a^*}.\end{cases}$$

Let $I = (i_1,\ldots,i_s)$ and $J = (j_1,\ldots,j_q)$ with $(I,J)\in\Gamma_W(S,T)$. If $\varphi,\psi\in\mathcal{C}_0^\infty(\mathbb{R}^N)$ have support in the unit ball, it follows from from part (a) of Lemma 9.4 that there is a function $\theta\in\mathcal{C}_0^\infty(\mathbb{R}^N)$, normalized with respect to the functions φ and ψ, so that

$$[\varphi]_I * [\psi]_J(\mathbf{x}) = 2^{-\sum_{j=1}^{\alpha+\beta} m_j Q_j^*}\,\theta(2^{-m_1}\hat{\circ}\,\overline{\mathbf{x}}_1,\ldots,2^{-m_{\alpha+\beta}}\hat{\circ}\,\overline{\mathbf{x}}_{\alpha+\beta}),$$

where Q_j^* is the homogeneous dimension of V_j, and

$$\begin{aligned} m_j &= i_r && \text{if } \overline{\mathbf{x}}_m = \mathbf{x}_{I_r^*}\\ m_j &= j_a && \text{if } \overline{\mathbf{x}}_m = \mathbf{x}_{J_a^*}.\end{aligned}$$

Thus the indices $\{m_1,\ldots,m_{\alpha+\beta}\}$ are some permutation of $\{r_1,\ldots,r_\alpha,a_1,\ldots,a_\beta\}$. Note that if $S^{**} = \{r_{\alpha+1},\ldots,r_s\}\subset\{1,\ldots,s\}$ and $T^{**} = \{a_{\beta+1},\ldots,a_q\}\subset\{1,\ldots,q\}$ are the free indices, they do not appear among $\{m_1,\ldots,m_{\alpha+\beta}\}$.

Now suppose that φ has cancellation in the variables $\{\mathbf{x}_{k_1},\ldots,\mathbf{x}_{k_s}\}$ and that ψ has cancellation in the variables $\{\mathbf{x}_{\ell_1},\ldots,\mathbf{x}_{\ell_q}\}$. In order to obtain additional estimates of the function θ, we will use part (b) of Lemma 9.4. Using the notation of Lemma 9.4, we have

$$\begin{aligned} C &= \{k_1,\ldots,k_s\}\subset\{1,\ldots,n\},\\ D &= \{\ell_1,\ldots,\ell_q\}\subset\{1,\ldots,n\}.\end{aligned}$$

The dilations are given by

$$\begin{aligned} \lambda_j &= 2^{i_r/e(k_r,j)} && \text{if } j\in I_r,\\ \mu_k &= 2^{j_a/e(\ell_a,j)} && \text{if } j\in J_a.\end{aligned}$$

The set $C_1 = \{k_r : \lambda_{k_r} < \mu_{k_r}\}$ is now the union of two disjoint subsets:

$$C_1' = \Big\{k_r : I_r^* = \emptyset\Big\} \qquad = \Big\{k_r : r \text{ is free}\Big\},$$
$$C_1'' = \Big\{k_r : I_r^* \neq \emptyset \text{ and } k_r \notin I_r^*\Big\} \quad = \Big\{k_r : r \text{ is fixed and } k_r \notin I_r^*\Big\}.$$

The same is true of the set $D_1 = \{\ell_a : \mu_{\ell_a} \leq \lambda_{\ell_a}\}$:

$$D_1' = \Big\{\ell_a : J_a^* = \emptyset\Big\} \qquad = \Big\{\ell_a : a \text{ is free}\Big\},$$
$$D_1'' = \Big\{\ell_a : J_a^* \neq \emptyset \text{ and } \ell_a \notin J_a^*\Big\} \quad = \Big\{\ell_a : a \text{ is fixed and } \ell_a \notin J_a^*\Big\}.$$

The sets C_2 and D_2 are given by

$$C_2 = \Big\{k_r : r \text{ is fixed and } k_r \in I_r^*\Big\},$$
$$D_2 = \Big\{\ell_a : a \text{ is fixed and } \ell_a \in J_a^*\Big\}.$$

Define maps

$$\sigma_1 : \Big\{1, \ldots, s\Big\} \to \Big\{1, \ldots, q\Big\} \text{ such that } k_r \in J_{\sigma_1(r)},$$
$$\sigma_2 : \Big\{1, \ldots, q\Big\} \to \Big\{1, \ldots, s\Big\} \text{ such that } \ell_a \in I_{\sigma_2(a)},$$

and note that

$$1 \leq r \leq s \text{ and } r \text{ is free} \quad \Longrightarrow \quad \sigma_1(r) \text{ is fixed},$$
$$1 \leq a \leq q \text{ and } a \text{ is free} \quad \Longrightarrow \quad \sigma_2(a) \text{ is fixed}.$$

Applying part (b) of Lemma 9.4 we have the following result.

LEMMA 9.13. *Let $(I, J) \in \Gamma(\mathbf{E}_{S,T,W})$, let φ, ψ be normalized bump functions, and suppose that φ has cancellation in the variables $\{\mathbf{x}_{k_1}, \ldots, \mathbf{x}_{k_s}\}$ and that ψ has cancellation in the variables $\{\mathbf{x}_{\ell_1}, \ldots \mathbf{x}_{\ell_q}\}$. Write $[\varphi]_I * [\psi]_J = [\theta]_M$ as in Lemma 9.3. There exists $\epsilon > 0$ depending only on the group structure so that θ can be written as a finite sum of terms of the form*

$$\prod_{k_r \in C_1'} 2^{-\epsilon G_1(i_r)} \prod_{k_r \in C_1''} 2^{-\epsilon G_1(i_r)} \prod_{\ell_a \in D_1'} 2^{-\epsilon H_1(j_a)} \times$$
$$\prod_{\ell_a \in D_1''} 2^{-\epsilon H_1(a)} \prod_{k_r \in C_2'} 2^{-\epsilon G_2(r)} \prod_{\ell_a \in D_2'} 2^{-\epsilon H_2(j_a)} \, \theta_{C_2'', D_2''}$$

where the factors are defined as follows.

a) *C_2 and D_2 are each partitioned into two disjoint subsets:*

$$C_2 = C_2' \cup C_2'', \quad C_2' \cap C_2'' = \emptyset,$$
$$D_2 = D_2' \cup D_2'', \quad D_2' \cap D_2'' = \emptyset.$$

b) *The functions $\{\theta_{C_2'', D_2''}\} \subset \mathcal{C}_0^\infty(\mathbb{R}^n)$ are normalized relative to φ and ψ.*

c) *The function $\theta_{C_2'', D_2''}$ has cancellation in each variable $\overline{\mathbf{x}}_m = \mathbf{x}_{I_r^*}$ if $r \in C_2''$.*

d) *The function $\theta_{C_2'', D_2''}$ has cancellation in each variable $\overline{\mathbf{x}}_m = \mathbf{x}_{J_a^*}$ if $a \in D_2''$.*

e) *The expressions $G_j(r)$ and $H_j(a)$ are given by:*

$$G_1(i_r) = \min\Big\{\frac{j_{\sigma_1(r)}}{e(\ell_{\sigma_1(r)}, k_r)} - i_r,\ i_{r+1} - i_r\Big\};$$

$$H_1(j_a) = \min\Big\{\frac{k_{\sigma_2(a)}}{e(k_{\sigma_2(a)}, \ell_a)} - j_a,\ j_{a+1} - j_a\Big\};$$

$$G_2(i_r) = i_{r+1} - i_r;$$

$$H_2(j_a) = j_{a+1} - j_a.$$

The meaning of this result is as follows. The original functions φ^I or ψ^J have cancellation in variables whose indices belong to $C_1 \cup C_2$ and to $D_1 \cup D_2$. Then with $\varphi^I * \psi^J = [\theta^{I,J}]_M$, the function $\theta^{I,J}$ can be written as a sum of terms in which a normalized bump function $\Theta^{I,J}_{C_1',D_1'}$ are multiplied by certain small factors.

A) If $r \in \{1,\dots,s\}$ or $a \in \{1,\dots,q\}$ is a free variable then $r \in C_1'$ or $a \in D_1'$. If r is free, the scale of $[\varphi^I]_I$ in the coordinate $\mathbf{x}_{k_r}$ must be smaller than the scale of $[\psi^J]_J$ in $\mathbf{x}_{k_r}$, which is $j_{\sigma_1(r)}/e(\ell_{\sigma_1(r)}, k_r)$. Thus there is a gain, either of order $-\epsilon[j_{\sigma_1(r)}/e(\ell_{\sigma_1(r)}, k_r) - i_r]$ or of order $-\epsilon[i_{r+1} - i_r]$, due either to integration by parts or to replacing the derivative $\partial_{\mathbf{x}_{k_r}}$ with the corresponding invariant differential operator. This leads to the term $2^{-\epsilon G_1(r)}$, and there is a similar explanation for the term $2^{-\epsilon H_1(a)}$. In particular, there is an gain associated with all free indices.

B) If $r \in \{1,\dots,s\}$ or $a \in \{1,\dots,q\}$ is a fixed variable, then $r \in C_1'' \cup C_2$ or $a \in D_1'' \cup D_2$.

   a) For indices $r \in C_1''$, $k_r \notin I_r^*$ and so the scale of $[\varphi^I]_I$ in the coordinate $\mathbf{x}_{k_r}$ must be smaller than the scale of $[\psi^J]_J$ in $\mathbf{x}_{k_r}$, which is $j_{\sigma_1(r)}/e(\ell_{\sigma_1(r)}, k_r)$. Thus there is a gain, either of order

   $$-\epsilon[j_{\sigma_1(r)}/e(\ell_{\sigma_1(r)}, k_r) - i_r]$$

   or of order $-\epsilon[i_{r+1} - i_r]$, due either to integration by parts or to replacing the derivative $\partial_{\mathbf{x}_{k_r}}$ with the corresponding invariant differential operator. This leads to the term $2^{-\epsilon G_1(r)}$ and there is a similar explanation for the term $2^{-\epsilon H_1(a)}$.

   b) For an index $r \in C_2$, it may happen that the cancelation of $[\varphi^I]_I$ in the variable $\mathbf{x}_{k_r}$ leads to cancellation in this variable of the convolution $[\varphi^I]_I * [\psi^J]_J$. We let C_2'' be the elements of C_2 where this happens and C_2' be the complementary set. Then $\theta^{I,J}_{C_2'',D_2''}$ does have cancellation in the variables $\mathbf{x}_{I_r^*}$ for $r \in C_2''$. In the same way,$\theta^{I,J}_{C_2'',D_2''}$ does have cancellation in the variables $\mathbf{x}^*_{J_a}$ for $a \in D_2''$.

   c) For indices $r \in C_2' = C_2 \setminus C_2''$ or $a \in D_2' = D_2 \setminus D_2''$, since cancellation does not persist in the convolution, there is an exponential gain of order $-\epsilon[i_{r+1} - i_r|$ or $-\epsilon[j_{a+1} - j_a|$ due replacing an ordinary derivative by a left- or right-invariant operator. This leads to the terms $2^{-\epsilon G_2(r)}$ and $2^{-\epsilon H_2(a)}$.

REMARKS 9.14. Let $(I,J) \in \Gamma(\mathbf{E}_{S,T,W})$.

1) If $k_r \in C_1'$ then $i_r \le j_{\sigma_1(r)} e(\ell_{\sigma_1(r)}, k_r)^{-1}$ and $\sigma_1(r)$ is a fixed index. Similarly, if $\ell_a \in D_1'$ then $j_a \le k_{\sigma_2(a)} e(k_{\sigma_2(a)}, \ell_a)^{-1}$ and $\sigma_2(a)$ is a fixed index.

2) If $k_r \in C_1''$ then $\sigma_1(r)$ is also fixed, and

$$\frac{i_r}{j_{\sigma_1(r)}} \geq \tau_W(J_{\sigma_1(r)}, I_r)^{-1} \geq \Lambda_W(J_{\sigma_1(r)}, I_r)^{-1}$$
$$= \frac{e(k_r, v_{m_{r,\sigma_1(r)}})}{e(\ell_{\sigma_1(r)}, v_{m_{r,\sigma_1(r)}})} \geq e(\ell_{\sigma_1(r)}, k_r)^{-1}$$

where, since $J^*_{\sigma(r)} \cap I_r \neq \emptyset$, the second inequality follows from the definition of τ_W. The last inequality follows from the basic hypothesis. Thus we have $i_r - e(\ell_{\sigma_1(r)}, k_r)^{-1} j_{\sigma_1(r)} \leq 0$. Similarly, if $j_a \in D_1''$, it follows that $j_a - e(k_{\sigma_2(a)}, \ell_a)^{-1} i_{\sigma_2(a)} \leq 0$.

9.10. Proof of Theorem 9.2

We have seen on page 73 that Theorem 9.2 follows from Lemma 9.3 which deals with sums of convolutions $[\varphi^I]_I * [\psi^J]_J$ where $(I, J) \in \Gamma_{\mathbb{Z}}(\mathbf{E}_S) \times \Gamma_{\mathbb{Z}}(\mathbf{E}_T)$. In section 9.4 we partitioned $\Gamma_{\mathbb{Z}}(\mathbf{E}_S) \times \Gamma_{\mathbb{Z}}(\mathbf{E}_T)$ into disjoint sets $\Gamma_W(S, T)$ for $W \in \mathcal{W}(S, T)$. Thus the proof of Lemma 9.3, and hence Theorem 9.2, follows from the following result.

LEMMA 9.15. *If $F \subset \Gamma_W(S, T)$ is any finite set, then*

$$\mathcal{H}_F = \sum_{(S,T)\in F} [\varphi^I]_I * [\psi^J]_J \in \mathcal{P}_0(\mathbf{E}_{S,T,W})$$

with constants independent of the finite set F. Moreover, $\lim_{F \nearrow \Gamma_W(S,T)} \mathcal{H}_F = \mathcal{H}_{S,T,W} \in \mathcal{P}_0(\mathbf{E}_{S,T,W})$ with convergence in the sense of distributions. In particular, it follows from Lemma 9.12 that $\mathcal{H}_{S,T,W} \in \mathcal{P}_0(\mathbf{E})$.

PROOF. Recall from page 72 that

$$\sum_{(I,J)\in\Gamma_W(S,T)} [\varphi^I]_I * [\psi^J]_J = \sum_{M\in\mathfrak{M}(S,T,W)} \Big[\sum_{(I,J)\in F(M)} \theta^{I,J} \Big]_M = \sum_{M\in\mathfrak{M}(S,T,W)} [\Theta^M]_M.$$

We first show that for each $M \in \mathfrak{M}(S, T, W)$, the sum $\sum_{(I,J)\in F(M)} \theta^{I,J}$ converges to a normalized bump function. Given $M = (m_1, \ldots, m_{\alpha+\beta})$, this determines the indices i_r and j_a for which r and a are fixed; *i.e.* $r \in \{r_1, \ldots, r_\alpha\}$ and $a \in \{a_1, \ldots, a_\beta\}$. Thus i_r and j_a belong to the sets C_1'', D_1'', C_2, D_2. We must sum over the remaining indices corresponding to free variables r and a for which $k_r \in C_1'$ and $\ell_a \in D_1'$. We write the free variables as $\{r_{\alpha+1}, \ldots, r_s\}$ and $\{a_{\beta+1}, \ldots, a_q\}$. Using Lemma 9.13 it follows that $\sum_{(I,J)\in F(M)} \theta^{I,J}$ is a finite sum of terms (indexed by subsets $C_2'' \subset C_2$ and $D_2'' \subset D_2$) of the form

$$\prod_{k_r\in C_1''} 2^{-\epsilon G_1(r)} \prod_{\ell_a\in D_1''} 2^{-\epsilon H_1(a)} \prod_{k_r\in C_2'} 2^{-\epsilon G_2(r)} \prod_{\ell_a\in D_2'} 2^{-\epsilon H_2(a)}$$
$$\Big[\sum_{\substack{(i_{r_{\alpha+1}},\ldots,i_{r_s})\in C_1' \\ (j_{a_{\beta+1}},\ldots,j_{a_q})\in D_1'}} \prod_{v=\alpha+1}^{s} 2^{-\epsilon G_1(r_v)} \prod_{v=\beta+1}^{q} 2^{-\epsilon H_1(a_v)} \theta_{C_2'',D_2''} \Big]$$

where each function $\theta_{C_2'',D_2''}$ has cancellation in the variables $\overline{\mathbf{x}}_m = \mathbf{x}_{I_r^*}$ if $r \in C_2''$ and $\overline{\mathbf{x}}_m = \mathbf{x}_{J_a^*}$ if $a \in D_2''$.

We have

$$\sum_{\substack{(i_{r_{\alpha+1}},\dots,i_{r_s})\in C_1' \\ (j_{a_{\beta+1}},\dots,j_{a_q})\in D_1'}} \prod_{v=\alpha+1}^{s} 2^{-\epsilon G_1(r_v)} \prod_{v=\beta+1}^{q} 2^{-\epsilon H_1(a_v)} \,\theta_{C_2'',D_2''}$$

$$\leq \prod_{v=\alpha+1}^{s} \Bigg[\sum_{i_{r_v}\leq j_{\sigma_1(r_v)} e(\ell_{\sigma_1(r_v)},k_{r_v})} 2^{-\epsilon[j_{\sigma_1(r_v)}e(\ell_{\sigma_1(r_v)},k_{r_v})-i_{r_v}]}\Bigg]$$

$$\prod_{v=\beta+1}^{q} \Bigg[\sum_{j_{a_v}\leq i_{\sigma_2(a_v)} e(k_{\sigma_2(a_v)},\ell_{a_v})} 2^{-\epsilon[i_{\sigma_2(a_v)}e(k_{\sigma_2(a_v)},\ell_{a_v})-j_{a_v}]}\Bigg] < \infty.$$

It follows that $\sum_{(I,J)\in F(M)} \theta^{I,J}$ is a finite sum of terms of the form

$$\prod_{k_r\in C_1''} 2^{-\epsilon G_1(r)} \prod_{\ell_a\in D_1''} 2^{-\epsilon H_1(a)} \prod_{k_r\in C_2'} 2^{-\epsilon G_2(r)} \prod_{\ell_a\in D_2'} 2^{-\epsilon H_2(a)} \,\theta^{I,J}_{M,C_2'',D_2''}$$

where $\theta^{I,J}_{M,C_2'',D_2''}$ is a normalized bump function. It remains to show that

$$\sum_{M\in\mathfrak{M}(S,T,W)} \prod_{k_r\in C_1''} 2^{-\epsilon G_1(r)} \prod_{\ell_a\in D_1''} 2^{-\epsilon H_1(a)} \prod_{k_r\in C_2'} 2^{-\epsilon G_2(r)} \prod_{\ell_a\in D_2'} 2^{-\epsilon H_2(a)} \,\theta^{I,J}_{M,C_2'',D_2''}$$

converges in the sense of distributions to an element of $\mathcal{P}_0(\mathbf{E}_{S,T,W})$. However, this follows from Theorem 5.13 once we observe that each term in the sum is a function with weak cancellation with parameter $\epsilon > 0$ relative to the multi-index M. This completes the proof. □

CHAPTER 10

Convolution of Calderón-Zygmund kernels

In this chapter we study the convolution of two or more compactly supported Calderón-Zygmund kernels with different homogeneities, and show that the result belongs to an appropriate class $\mathcal{P}_0(\mathbf{E})$.

10.1. Calderón-Zygmund kernels

Recalling the notion of Calderón-Zygmund kernel given in Definition 7.1, we define the corresponding class of kernels with rapid decay at infinity.

DEFINITION 10.1. $\mathcal{CZ}_{\mathbf{a},0}$ is the space of tempered distributions $\mathcal{K}$ on $\mathbb{R}^N$ such that

(a) away from the origin, $\mathcal{K}$ is given by integration against a smooth function K satisfying the differential inequalities
$$\big|\partial^\gamma K(\mathbf{x})\big| \leq C_{\gamma,M} N_{\mathbf{a}}(\mathbf{x})^{-Q_{\mathbf{a}}-[\![\gamma]\!]_{\mathbf{a}}}\big(1+N_{\mathbf{a}}(\mathbf{x})\big)^{-M},$$
for every γ and M;

(b) there is a constant $C>0$ so that for any normalized bump function $\psi \in \mathcal{C}_0^\infty(\mathbb{R}^N)$ with support in the unit ball and for any $R>0$, $\big|\langle \mathcal{K}, \psi_R\rangle\big| \leq C$, where $\psi_R(\mathbf{x}) = \psi(R \cdot_{\mathbf{a}} \mathbf{x})$.

It follows from the results of Section 4 that, for distributions $\mathcal{K} \in \mathcal{CZ}_{\mathbf{a},0}$, the corresponding multipliers $m = \widehat{\mathcal{K}}$ are characterized by the differential inequalities
$$\big|\partial^\gamma m(\boldsymbol{\xi})\big| \leq C_\gamma\big[1+N_{\mathbf{a}}(\boldsymbol{\xi})\big]^{-[\![\gamma]\!]_{\mathbf{a}}}.$$
These differential inequalities characterize the multipliers belonging to the class $\mathcal{M}_\infty(\mathbf{E_a})$ where $\mathbf{E_a}$ is the matrix in (7.3). It follows that, if $\mathcal{K} \in \mathcal{CZ}_{\mathbf{a},0}$, then $\mathcal{K} \in \mathcal{P}_0(\mathbf{E_a})$. Moreover, the matrix $\mathbf{E_a}$ is doubly monotone if and only if $a_1 \geq a_2 \geq \cdots \geq a_n$. Hence this is the compatibility condition with an underlying nilpotent Lie group structure in the sense of Section 8.1.

Finally, it follows from our decomposition result that if $\mathcal{K} \subset \mathcal{P}_0(\mathbf{E_a})$, there exists $\Phi \in \mathcal{S}(\mathbb{R}^N)$ and a uniformly bounded family $\{\varphi^j\} \subset \mathcal{C}_0^\infty(\mathbb{R}^N)$ such that $\int_{\mathbb{R}^N} \varphi^j(\mathbf{x})\,d\mathbf{x} = 0$ for all j, and, in the sense of distributions,

(10.1)
$$\mathcal{K}(\mathbf{x}) = \sum_{j=-\infty}^{0} [\varphi^j]_j(\mathbf{x}) + \Phi(\mathbf{x}) = \sum_{j=-\infty}^{0} 2^{-jQ_a}\,\varphi^j(2^{-ja_1}\cdot \mathbf{x}_1,\ldots,2^{\,ja_n}\cdot \mathbf{x}_n) + \Phi(\mathbf{x}).$$

10.2. A general convolution theorem

Let $\mathcal{K}_1,\ldots,\mathcal{K}_p$ be Calderón-Zygmund kernels on a homogeneous nilpotent Lie group $G \cong \mathbb{R}^N = \mathbb{R}^{C_1}\oplus\cdots\oplus\mathbb{R}^{C_n}$. Suppose that the kernel $\mathcal{K}_\ell$ is in the class $\mathcal{CZ}_{\mathbf{a}_\ell,0}$ associated with the dilations $\lambda \cdot_{\mathbf{a}_\ell} \mathbf{x} = \big(\lambda^{a_1^\ell}\cdot \mathbf{x}_1,\ldots,\lambda^{a_n^\ell}\cdot \mathbf{x}_n\big)$. If G is not abelian, we assume that (8.7) holds, as well as the compatibility condition $a_j^\ell \geq a_{j+1}^\ell$ for all j, ℓ.

THEOREM 10.2.

(a) *The convolution* $\mathcal{K}_1 * \cdots * \mathcal{K}_p$ *belongs to the class* $\mathcal{P}_0(\mathbf{E})$ *where the entries of the matrix are given by* $e(j,k) = \max_{1\leq \ell \leq p} \frac{a_j^\ell}{a_k^\ell}$.

(b) *Conversely, every* $\mathbf{E}$ *arises in this way. Precisely, let* $\mathbf{E} = \{e(j,k)\}$ *be any* $n \times n$ *matrix satisfying* (2.10). *There are* n *sequences* $\mathbf{a}_l$ *such that* $e(j,k) = \max_{1\leq \ell\leq n} \frac{a_j^\ell}{a_k^\ell}$. *If* $\mathbf{E}$ *is doubly monotone, each* $\mathbf{a}_l$ *can be taken non-increasing.*

PROOF. We have the identity $\mathcal{CZ}_{\mathbf{a}_\ell,0} = \mathcal{P}_0(\mathbf{E}_\ell)$, where $e_\ell(j,k) = \big(a_j^\ell\big)\big(a_k^\ell\big)^{-1}$. By Proposition 5.2, each $\mathcal{K}_\ell \in \mathcal{P}_0(\mathbf{E})$. Moreover, if $a_j^\ell \geq a_{j+1}^\ell$ for all j, ℓ, $\mathbf{E}$ is doubly monotone. Then part (a) follows from Theorems 9.1 and 9.2.

To establish (b), put $a_j^\ell = e(j,\ell)$. Since $\mathbf{E}$ is doubly monotone, $a_j^\ell = e(j,\ell) \geq e(j+1,\ell) = a_{j+1}^\ell$, so $a_1^\ell \geq a_2^\ell \geq \cdots \geq a_n^\ell$. For any $1 \leq j, \ell \leq n$ we have

$$\frac{a_j^\ell}{a_k^\ell} = \frac{e(j,\ell)}{e(k,\ell)} \leq \frac{e(j,k)e(k,\ell)}{e(k,\ell)} = e(j,k)$$

so $\max_{1\leq \ell\leq m} \big(a_j^\ell\big)\big(a_k^\ell\big)^{-1} \leq e(j,k)$. On the other hand

$$e(j,k) = \frac{e(j,k)}{e(k,k)} \leq \max_{1\leq \ell\leq n} \frac{e(j,\ell)}{e(k,\ell)} = \max_{1\leq \ell\leq n} \frac{a_j^\ell}{a_k^\ell}.$$

This completes the proof. □

10.3. Convolution of two Calderón-Zygmund kernels

Part (a) of Theorem 10.2 determines the minimal class $\mathcal{P}_0(\mathbf{E})$ containing a finite family of Calderón-Zygmund classes $\mathcal{CZ}_{\mathbf{a}_\ell,0}$ and part (b) shows that every class $\mathcal{P}_0(\mathbf{E})$ arises in this way. It is possible, however, that a convolution

$$\mathcal{K}_1 * \mathcal{K}_2 * \cdots * \mathcal{K}_p,$$

with $\mathcal{K}_j \in \mathcal{CZ}_{\mathbf{a}_j,0}$, is contained in a proper subalgebra of the class $\mathcal{P}_0(\mathbf{E})$ which appears in Theorem 10.2 (a).

We show that such a situation occurs with two Calderón-Zygmund classes having different homogeneities. Let $\mathbf{a} = (a_1, \ldots, a_n)$ and $\mathbf{b} = (b_1, \ldots, b_n)$ with $a_j \geq a_{j+1}$ and $b_j \geq b_{j+1}$ for all j, and set

$$\lambda \cdot_{\mathbf{a}} \mathbf{x} = \big(\lambda^{a_1}\mathbf{x}_1, \ldots, \lambda^{a_n}\mathbf{x}_n\big),$$
$$\lambda \cdot_{\mathbf{b}} \mathbf{x} = \big(\lambda^{b_1}\mathbf{x}_1, \ldots, \lambda^{b_n}\mathbf{x}_n\big).$$

Let $\mathcal{K} \in \mathcal{P}_0(\mathbf{E_a})$ and $\mathcal{L} \in \mathcal{P}_0(\mathbf{E_b})$. If we apply part (a) of Theorem 10.2, we obtain that $\mathcal{K} * \mathcal{L} \in \mathcal{P}_0(\mathbf{E})$, where the entries of $\mathbf{E}$ are

$$e(j,k) = \max\Big\{\frac{a_j}{a_k}, \frac{b_j}{b_k}\Big\}. \tag{10.2}$$

Notice that

$$e(j,k)e(k,j) = \max\Big\{\frac{b_j/a_j}{b_k/a_k}, \frac{b_k/a_k}{b_j/a_j}\Big\},$$

so that the reduced rank of $\mathbf{E}$ can be as large as n. We will show that $\mathcal{K}*\mathcal{L}$ belongs to the sum of $n-1$ classes $\mathcal{P}_0(\mathbf{E}_m)$ where each $\mathbf{E}_m$ has reduced rank equal to 2.

As in (10.1), write

$$\begin{aligned}\mathcal{K}(\mathbf{x}) &= \sum_{j=-\infty}^{0} [\varphi^j]_j(\mathbf{x}) + \Phi(\mathbf{x}) \\ &= \sum_{j=-\infty}^{0} 2^{-jQ_{\mathbf{a}}}\varphi^j(2^{-ja_1}\cdot\mathbf{x}_1,\ldots,2^{-ja_n}\cdot\mathbf{x}_n) + \Phi(\mathbf{x}), \\ \mathcal{L}(\mathbf{x}) &= \sum_{k=-\infty}^{0} [\psi^k]_k(\mathbf{x}) + \Psi(\mathbf{x}) \\ &= \sum_{k=-\infty}^{0} 2^{-kQ_{\mathbf{b}}}\psi^k(2^{-kb_1}\cdot\mathbf{x}_1,\ldots,2^{-kb_n}\cdot\mathbf{x}_n) + \Psi(\mathbf{x}).\end{aligned}$$

Since $\mathcal{K}*\Psi,\ \Phi*\mathcal{L},\ \Phi*\Psi\in\mathcal{S}(\mathbb{R}^N)$, we only need to analyze the double sum

$$\sum_{j=-\infty}^{0}\sum_{k=-\infty}^{0}[\varphi^j]_j*[\psi^k]_k. \tag{10.3}$$

Let σ be the permutation of $\{1,\ldots,n\}$ so that $\frac{b_{\sigma(1)}}{a_{\sigma(1)}}\leq\frac{b_{\sigma(2)}}{a_{\sigma(2)}}\leq\cdots\leq\frac{b_{\sigma(n)}}{a_{\sigma(n)}}$. For simplicity, we will suppose that all the inequalities are strict[1].

We split the double sum in (10.3) into $n+1$ two-parameter sums. Set

$$\mathcal{N}_m = \sum_{(j,k)\in\Gamma_m}[\varphi^j]_j*[\psi^k]_k,\quad 0\leq m\leq n, \tag{10.4}$$

where

(10.5)

$$\begin{aligned}\Gamma_0 &= \Big\{(j,k)\in\mathbb{Z}\times\mathbb{Z}: j<0,\quad k<0,\quad \frac{j}{k}\leq\frac{b_{\sigma(1)}}{a_{\sigma(1)}}\Big\}, \\ \Gamma_m &= \Big\{(j,k)\in\mathbb{Z}\times\mathbb{Z}: j<0,\quad k<0,\quad \frac{b_{\sigma(m)}}{a_{\sigma(m)}}<\frac{j}{k}\leq\frac{b_{\sigma(m+1)}}{a_{\sigma(m+1)}}\Big\},\ 1\leq m\leq n-1, \\ \Gamma_n &= \Big\{(j,k)\in\mathbb{Z}\times\mathbb{Z}: j<0,\quad k<0,\quad \frac{b_{\sigma(n)}}{a_{\sigma(n)}}<\frac{j}{k}\Big\}.\end{aligned}$$

Then clearly $\sum_{j=-\infty}^{0}\sum_{k=-\infty}^{0}[\varphi^j]_j*[\psi^k]_k=\sum_{m=0}^{n}\mathcal{N}_m$.

LEMMA 10.3. *For $0\leq m\leq n$ let $A_m=\{\sigma(1),\ldots,\sigma(m)\}$ and $B_m=\{\sigma(m+1),\ldots,\sigma(n)\}$. Then the distribution $\mathcal{N}_m\in\mathcal{P}_0(\mathbf{E}_m)$ where $\mathbf{E}_m=\{e_m(j,k)\}$ is the*

[1]The proof below shows that this restriction does not affect the validity of the final result.

$n \times n$ matrix whose entries are given by

$$\begin{aligned}
e_m(j,k) &= \frac{b_j}{b_k} && \textit{if } j \in A_m \textit{ and } k \in A_m,\\
e_m(j,k) &= \frac{b_j a_{\sigma(m)}}{a_k b_{\sigma(m)}} && \textit{if } j \in A_m \textit{ and } k \in B_m,\\
e_m(j,k) &= \frac{a_j b_{\sigma(m+1)}}{b_k a_{\sigma(m+1)}} && \textit{if } j \in B_m \textit{ and } k \in A_m,\\
e_m(j,k) &= \frac{a_j}{a_k} && \textit{if } j \in B_m \textit{ and } k \in B_m.
\end{aligned}$$

PROOF. We use the fact that

$$[\varphi^j]_j * [\psi^k]_k(\mathbf{x}) = 2^{-\sum_{p=1}^n l_p Q_p} \theta^{j,k}(2^{-l_1}\cdot \mathbf{x}_1, \dots, 2^{-l_n}\cdot \mathbf{x}_n)$$

where $l_p = ja_p \vee kb_p$. Now if $(j,k) \in \Gamma_m$ then

$$l_p = \begin{cases} kb_p & \text{if } p \in \{\sigma(1), \dots, \sigma(m)\} = A_m, \\ \\ ja_p & \text{if } p \in \{\sigma(m+1), \dots, \sigma(n)\} = B_m. \end{cases} \tag{10.6}$$

If $L(j,k)$ is the multi-index with components l_p as above, one easily verifies that

$$(j,k) \in \Gamma_m \Longrightarrow L(j,k) \in \Gamma(\mathbf{E}_m).$$

There are four cases to cosider.

(1) If $p, r \in \{\sigma(1), \dots, \sigma(m)\}$ then $l_p = kb_p$ and $l_r = kb_r$, and so $e_m(p,r)l_r = \frac{b_p}{b_r}\, kb_r = k\, b_p = l_p$, which shows $e_m(p,r)l_r \leq l_p$ as required.

(2) If $p, r \in \{\sigma(m+1), \dots, \sigma(n)\}$ then $l_p = ja_p$ and $l_r = ja_r$, and so $e_m(p,r)l_r = \frac{a_p}{a_r}\, ja_r = j\, a_p = l_p$.

(3) If $p \in \{\sigma(1), \dots, \sigma(m)\}$ and $r \in \{\sigma(m+1), \dots, \sigma(n)\}$ then $l_p = kb_p$ and $l_r = ja_r$. Then

$$e_m(p,r)l_r = \frac{b_p a_{\sigma(m)}}{a_r b_{\sigma(m)}}\, j\, a_r = \frac{ja_{\sigma(m)}}{kb_{\sigma(m)}}\, kb_p = \frac{ja_{\sigma(m)}}{kb_{\sigma(m)}}\, l_p \leq l_p.$$

since $\frac{ja_{\sigma(m)}}{kb_{\sigma(m)}} \geq 1$ and $l_p < 0$.

(4) If $p \in \{\sigma(m+1), \dots, \sigma(n)\}$ and $r \in \{\sigma(1), \dots, \sigma(m)\}$ then $l_p = ja_p$ and $l_r = kb_r$. Then

$$e_m(p,r)l_r = \frac{a_p b_{\sigma(m+1)}}{b_r a_{\sigma(m+1)}}\, k\, b_r = \frac{ka_{\sigma(m+1)}}{jb_{\sigma(m+1)}}\, ja_p = \frac{k\, b_{\sigma(m+1)}}{j\, a_{\sigma(m+1)}}\, l_p \leq l_p.$$

since $\frac{k\, b_{\sigma(m+1)}}{j\, a_{\sigma(m+1)}} \geq 1$ and $l_p < 0$.

Consider first the case $1 \leq m \leq n-1$, where A_m and B_m are both nonempty. Then the map $(j,k) \in \Gamma_m \longmapsto L(j,k)$ is injective. Since each term $[\varphi^j]_j * [\psi^k]_k$ has weak cancellation, it follows[2] from Lemma 5.13 that the series (10.4) convergens to a kernel in the class $\mathcal{P}_0(\mathbf{E}_m)$.

[2]The fact that the components l_p are not integers is irrelevant as long as they belong to a fixed lattice.

Suppose now that $m = 0$ (the case $m = n$ can be treated in the same way). In this case the matrix $\mathbf{E}_0$ has rank 1 and the statement is that $\mathcal{N}_0 \in \mathcal{CZ}_{\mathbf{a},0}$. Notice that $L(j,k) = j\mathbf{a}$ only depends on j. Hence for each j we must verify that

$$\sum_{k \leq \frac{a_{\sigma(1)}}{b_{\sigma(1)}} j} [\varphi^j]_j * [\psi^k]_k = [\widetilde{\varphi}^j]_j$$

with uniformly bounded functions $\widetilde{\varphi}^j$ with mean value zero (the scaling by j being as in (10.1)). It follows from Lemma 9.4 that each term in the sum is a normalized bump function times a factor that decreases exponentially with k. The mean value zero property is obvious. □

REMARKS 10.4.

1. Recall that

$$\overbrace{\frac{b_{\sigma(1)}}{a_{\sigma(1)}} < \frac{b_{\sigma(2)}}{a_{\sigma(2)}} < \cdots < \frac{b_{\sigma(m)}}{a_{\sigma(m)}}}^{\sigma\text{'s belonging to } A_m} < \overbrace{\frac{b_{\sigma(m+1)}}{a_{\sigma(m+1)}} < \cdots < \frac{b_{\sigma(n-1)}}{a_{\sigma(n-1)}} < \frac{b_{\sigma(n)}}{a_{\sigma(n)}}}^{\sigma\text{'s belonging to } B_m},$$

   and that
   $a_1 \geq a_2 \geq \cdots \geq a_n$, $b_1 \geq b_2 \geq \cdots \geq b_n$. We claim that $e(j+1,j) \leq 1$ for $1 \leq j \leq n-1$.
   - If $j, j+1 \in A_m$, then since $b_{j+1} \leq b_j$,
   $$e(j+1,j) = \frac{b_{j+1}}{b_j} \leq \frac{b_j}{b_j} = 1.;$$
   - If $j \in A_m$ and $j+1 \in B_m$ then $\frac{a_{j+1}}{b_{j+1}} \leq \frac{a_{\sigma(m+1)}}{b_{\sigma(m+1)}} \leq \frac{a_{\sigma(m)}}{b_{\sigma(m)}}$ and so
   $$e(j+1,j) = \frac{a_{j+1}b_{\sigma(m)}}{b_j a_{\sigma(m)}} \leq \frac{a_{j+1}b_{\sigma(m)}}{b_{j+1}a_{\sigma(m)}} \leq 1.$$
   - If $j \in B_m$ and $j+1 \in A_m$ then $\frac{b_{j+1}}{a_{j+1}} \leq \frac{b_{\sigma(m)}}{a_{\sigma(m)}}$ and so
   $$e(j+1,j) = \frac{b_{j+1}a_{\sigma(m+1)}}{a_j b_{\sigma(m+1)}} \leq \frac{b_{j+1}a_{\sigma(m+1)}}{a_{j+1}b_{\sigma(m+1)}} \leq \frac{b_{\sigma(m)}a_{\sigma(m+1)}}{a_{\sigma(m)}b_{\sigma(m+1)}} \leq 1.$$
   - If $j, j+1 \in B_m$, then since $a_{j+1} \leq a_j$,
   $$e(j+1,j) = \frac{a_{j+1}}{a_j} \leq 1.$$

   It now follows from Proposition 8.3 that the matrix $\mathbf{E}_m$ is doubly monotone.

2. For $m = 0, n$, $\mathbf{E}_m$ has rank one and $\mathcal{N}_0 \in \mathcal{CZ}_{\mathbf{a},0}$, $\mathcal{N}_n \in \mathcal{CZ}_{\mathbf{b},0}$. For $1 \leq m \leq n-1$, we can also think of the distribution $\mathcal{N}_m$ on the coarser decomposition
$$\mathbb{R}^N = \big(\oplus_{j\in A_m} \mathbb{R}^{C_j}\big) + \big(\oplus_{j\in B_m} \mathbb{R}^{C_j}\big).$$

with exponents, in the notation of Section 5.3, $\boldsymbol{\alpha}_m = (\alpha_{m,1}, \ldots, \alpha_{m,n})$ given by

$$\alpha_{m,j} = \begin{cases} \frac{1}{b_j} & \text{if } j \in A_m \\ \frac{1}{a_j} & \text{if } j \in B_m. \end{cases}$$

$$\mathbf{E}^\flat_m = \begin{bmatrix} 1 & \frac{a_{\sigma(m+1)}}{b_{\sigma(m+1)}} \\ \frac{b_{\sigma(m)}}{a_{\sigma(m)}} & 1 \end{bmatrix}.$$

(Note that the product of the off-diagonal elements is at least 1.)

3. The proof of Lemma 10.3 shows that, if G is the additive group $\mathbb{R}^N$, the inclusion (10.7) holds without any monotonicity assumption on the a_j and the b_j and with the same definition of $\sigma(m)$.[3]

PROPOSITION 10.5. We have the inclusion

$$\mathcal{CZ}_{\mathbf{a},0} * \mathcal{CZ}_{\mathbf{b},0} \subseteq \sum_{m=1}^{n-1} \mathcal{P}_0(\mathbf{E}_m), \tag{10.7}$$

Moreover, the class $\sum_{m=1}^{n-1} \mathcal{P}_0(\mathbf{E}_m)$ is contained in $\mathcal{P}_0(\mathbf{E})$, where $\mathbf{E}$ is the matrix with entries (10.2), and is closed under convolution.

PROOF. It is easy to check using the definition in Lemma 10.3 that $e_0(j,k) \leq e_1(j,k)$ for all j,k, and it then follows from Proposition 5.2 that $\mathcal{N}_0 \in \mathcal{P}_0(\mathbf{E}_0) \subset \mathcal{P}_0(\mathbf{E}_1)$. Similarly $\mathcal{N}_n \in \mathcal{P}_0(\mathbf{E}_n) \subset \mathcal{P}_0(\mathbf{E}_{n-1})$. This gives the stated inclusion. By Proposition 5.2, the first part of the statement follows from the inequalities

$$e_m(j,k) \leq \max\left\{\frac{a_j}{a_k}, \frac{b_j}{b_k}\right\},$$

which can be easily verified. To prove the second part of the statement, observe that each $\mathcal{K} \in \sum_{m=1}^{n-1} \mathcal{P}_0(\mathbf{E}_m)$ decomposes as

$$\mathcal{K} = \sum_{m=1}^{n-1} \mathcal{N}_m + \sum_{l=1}^{n} \mathcal{M}_l,$$

where $\mathcal{N}_m \in \mathcal{P}_0(\mathbf{E}_m)$ has the form

$$\mathcal{N}_m = \sum_{(j,k)\in\Gamma_m} [\varphi^{j,k}]_{L(j,k)},$$

where the $\varphi^{j,k}$ are uniformly bounded and have cancellation in the variables $\mathbf{x}_{A(m)}$ and $\mathbf{x}_{B(m)}$, and $\mathcal{M}_l \in \mathcal{CZ}_{\boldsymbol{\gamma}_l,0}$, where $\boldsymbol{\gamma}_l = (\gamma^l_1, \ldots, \gamma^l_n)$ and

$$\gamma^l_m = \begin{cases} a_{\sigma(l)} b_m & \text{if } \frac{b_m}{a_m} \leq \frac{b_{\sigma(l)}}{a_{\sigma(l)}}, \\ b_{\sigma(l)} a_m & \text{if } \frac{b_m}{a_m} \geq \frac{b_{\sigma(l)}}{a_{\sigma(l)}}. \end{cases}$$

[3]This can also be proved more directly by decomposing the multiplier $\mu = \widehat{\mathcal{K}}\widehat{\mathcal{L}}$. Assuming, as we may, that the variables have been ordered so that $\sigma(m) = m$ for every m, μ can be decomposed as a sum $\mu = \sum_{m=1}^{n-1} \mu_m$, where $\mu_m \in \mathcal{M}_\infty(\mathbf{E}_m)$ and $\mathbb{B}^c \cap \operatorname{supp} \mu_m$ is contained in the set where

$$N_{\mathbf{b}}(\boldsymbol{\xi}_1, \ldots, \boldsymbol{\xi}_m)^{b_m} \leq A N_{\mathbf{a}}(\boldsymbol{\xi}_m, \ldots, \boldsymbol{\xi}_n)^{a_m},$$
$$N_{\mathbf{a}}(\xi_{m+1}, \ldots, \boldsymbol{\xi}_n)^{a_{m+1}} \leq A N_{\mathbf{b}}(\boldsymbol{\xi}_1, \ldots, \boldsymbol{\xi}_{m+1})^{b_{m+1}}.$$

Given a second kernel $\mathcal{K}' = \sum_{m=1}^{n-1} \mathcal{N}'_m + \sum_{l=1}^{n} \mathcal{M}'_l$, we consider the convolutions

$$\mathcal{N}_m * \mathcal{N}'_{m'}, \qquad \mathcal{N}_m * \mathcal{M}'_{l'}, \qquad \mathcal{M}_l * \mathcal{N}'_{m'}, \qquad \mathcal{M}_l * \mathcal{M}'_{l'}.$$

We prove that, assuming $m \le m'$, $\mathcal{N}_m * \mathcal{N}'_{m'} \in \sum_{p=m}^{m'} \mathcal{P}_0(\mathbf{E}_p)$. The proofs of the following statements are left to the reader:

$$\mathcal{N}_m * \mathcal{M}'_{l'} \in \begin{cases} \sum_{p=m}^{l'-1} \mathcal{P}_0(\mathbf{E}_p) & \text{if } m < l' \\ \sum_{p=l'}^{m} \mathcal{P}_0(\mathbf{E}_p) & \text{if } m \ge l'. \end{cases} \qquad \text{(similarly for } \mathcal{M}_l * \mathcal{N}'_{m'})$$

$$\mathcal{M}_l * \mathcal{M}'_{l'} \in \sum_{p=l}^{l'-1} \mathcal{P}_0(\mathbf{E}_p) \qquad (l < l').$$

Let

$$\mathcal{N}_m = \sum_{(j,k)\in\Gamma_m} [\varphi^{j,k}]_{L(j,k)}, \qquad \mathcal{N}'_{m'} = \sum_{(j',k')\in\Gamma_{m'}} [\psi^{j',k'}]_{L'(j',k')},$$

where the components of $L'(j',k')$ are defined by (10.6) with m' in place of m. We isolate a single term $[\varphi^{j,k}]_{L(j,k)} * [\psi^{j',k'}]_{L'(j',k')}$ in the convolution, which is scaled by the multi-index $L(j,k) \vee L'(j',k') = (l_p)_{1\le p\le n}$, where

$$L(j,k)_p = \begin{cases} kb_p & \text{if } p \in A_m \\ ja_p & \text{if } p \in B_m \end{cases} \qquad L'(j',k')_p = \begin{cases} k'b_p & \text{if } p \in A_{m'} \\ j'a_p & \text{if } p \in B_{m'}. \end{cases}$$

Consider first the case $k \le k'$.

If $p \in A_m \subset A_{m'}$, then $l_p = k'b_p$.

Assume now that $r, s \in \{m+1, \dots, m'\}$, $r < s$. Then $p = \sigma(r), q = \sigma(s) \in B_m \cap A_{m'}$. If $ja_p \ge k'b_p$, then

$$j \ge k'\frac{b_p}{a_p} \ge k'\frac{b_q}{a_q}. \tag{10.8}$$

If

$$m'' = \max\{r : k'b_{\sigma(r)} > ja_{\sigma(r)}\}, \tag{10.9}$$

$$l_p = \begin{cases} k'b_p & \text{if } p = \sigma(r) \text{ with } r \le m'' \\ ja_p & \text{if } p = \sigma(r) \text{ with } m'' < r \le m'. \end{cases}$$

Assume that R' is nonempty, i.e., that, for $q = \sigma(m')$, $k'b_q \le ja_q$. Since $(j',k') \in \Gamma_{m'}$, we have

$$j' < k'\frac{b_q}{a_q} \le j.$$

Therefore $l_p = ja_p$ for all $p \in B_{m'}$. To conclude,

$$l_p = \begin{cases} k'b_p & \text{if } p \in A_{m''} \\ ja_p & \text{if } p \in B_{m''}. \end{cases}$$

Notice that $(j,k') \in \Gamma_{m''}$ by (10.9). We then have

$$\sum_{\substack{(j,k)\in\Gamma_m\\(j',k')\in\Gamma_{m'}\\k\le k'}} [\varphi^{j,k}]_{L(j,k)} * [\psi^{j',k'}]_{L'(j',k')} = \sum_{m''=m}^{m'} \sum_{(j,k')\in\Gamma_{m''}} \sum_{\substack{(j,k)\in\Gamma_m\\(j',k')\in\Gamma_{m'}\\k\le k',\, j'\le j}} [\varphi^{j,k}]_{L(j,k)} * [\psi^{j',k'}]_{L'(j',k')}. \tag{10.10}$$

By Lemma 9.4, the terms in the innermost sum are uniformly bounded up to a factor which decays exponentially in j' and k and have weak cancellation in the variables $\mathbf{x}_{A_{m''}}, \mathbf{x}_{B_{m''}}$. So the sum in (10.10) belongs to $\sum_{m''=m}^{m'} \mathcal{P}_0(\mathbf{E}_{m''})$.

Consider now $k > k'$. Then $l_p = ka_p$ for $p \in A_m$. Since $(j,k) \in \Gamma_m$, for $p = \sigma(m+1)$ we have $ja_p \ge kb_p > k'b_p$. Hence $l_p = ja_p$. By (10.8), we obtain that $l_p = ja_p$ for all $p \in B_m \cap A_{m'}$.

In particular, since $(j',k') \in \Gamma_{m'}$, for $p = \sigma(m')$ we have $ja_p \ge k'b_p > j'a_p$ so that $j > j'$. Then $l_p = ja_p$ for all $p \in B_{m'}$. To conclude, $L(j,k) \vee L'(j',k') = L(j,k) \in \Gamma_m$ and

$$\sum_{\substack{(j,k)\in\Gamma_m\\(j',k')\in\Gamma_{m'}\\k>k'}} [\varphi^{j,k}]_{L(j,k)} * [\psi^{j',k'}]_{L'(j',k')} = \sum_{(j,k)\in\Gamma_m} \sum_{\substack{(j',k')\in\Gamma_{m'}\\k'<k,\, j'<j}} [\varphi^{j,k}]_{L(j,k)} * [\psi^{j',k'}]_{L'(j',k')}. \tag{10.11}$$

Repeating the same arguments used in the other case, we conclude that the sum in (10.11) belongs to $\mathcal{P}_0(\mathbf{E}_m)$. □

CHAPTER 11

Two-flag kernels and multipliers

In this Section we show the surprising fact, already mentioned in Section 1.2, that distributions which are simultaneously flag kernels for two opposite flags are in fact the elements of an appropriate class $\mathcal{P}(\mathbf{E})$. We will recall the definition of flag-kernel in Section 11.1, and define two-flag kernels in Section 11.2. We consider the special case of step-two flags (which is considerably simpler than the general case) in Section 11.3, and the general situation in Section 11.4. The key point is that if m is the Fourier transform of a compactly supported distribution satisfying the differential inequalities of flag kernels, then m actually satisfies improved estimates. This is established in Section 11.5.

11.1. Flag kernels and multipliers

We recall the definition of *flag kernel* introduced in [**NRS01**]. Let $\mathbb{R}^N = \mathbb{R}^{C_1} \oplus \cdots \oplus \mathbb{R}^{C_n}$, and use the notation of Section 2, so that n_j is a homogeneous norm on $\mathbb{R}^{C_j}$, which has homogeneous dimension Q_j. An *n-step flag* is a strictly increasing sequence of n subspaces of $\mathbb{R}^N$, each of which is a direct sum of the subspaces $\{\mathbb{R}^{C_1}, \ldots, \mathbb{R}^{C_n}\}$. In particular, denote by $\mathcal{F}$ the flag

$$\mathcal{F}: \quad (0) \subsetneq \mathbb{R}^{C_n} \subsetneq \mathbb{R}^{C_{n-1}} \oplus \mathbb{R}^{C_n} \subsetneq \cdots \subsetneq \mathbb{R}^{C_2} \oplus \cdots \oplus \mathbb{R}^{C_n} \subsetneq \mathbb{R}^N. \tag{11.1}$$

Given an n-tuple $\mathbf{a} = (a_1, \ldots, a_n)$ of positive real numbers, we consider the dilations

$$\lambda \cdot_{\mathbf{a}} \mathbf{x} = (\lambda^{a_1} \cdot \mathbf{x}_1, \ldots, \lambda^{a_n} \cdot \mathbf{x}_n) \tag{11.2}$$

defined in (7.2) and introduce on $\mathbb{R}^N$ the *partial norms*

$$N_j'(\mathbf{x}) = n_1(\mathbf{x}_1)^{1/a_1} + \cdots + n_j(\mathbf{x}_j)^{1/a_j}, \tag{11.3}$$

so that N_j' does *not* involve the variables $\mathbf{x}_{j+1}, \ldots, \mathbf{x}_n$. Recall that if

$$L = \{l_1, \ldots, l_r\} \subset \{1, \ldots, n\}$$

with $l_1 < \cdots < l_r$, we denote by $\mathbb{R}^L$ the space $\mathbb{R}^{C_{l_1}} \oplus \cdots \oplus \mathbb{R}^{C_{l_r}}$. The "quotient flag" of $\mathcal{F}$ on $\mathbb{R}^L$, denoted by $\mathcal{F}_L$, is then

$$(0) \subsetneq \mathbb{R}^{C_{l_r}} \subsetneq \mathbb{R}^{C_{l_{r-1}}} \oplus \mathbb{R}^{C_{l_r}} \subsetneq \cdots \subsetneq \mathbb{R}^{C_{l_2}} \oplus \cdots \mathbb{R}^{C_{l_r}} \subsetneq \mathbb{R}^L \tag{11.4}$$

and the corresponding partial norms are given for $1 \leq j \leq r$ by

$$N_{L,j}'(\mathbf{x}_{l_1}, \ldots, \mathbf{x}_{l_j}) = n_{l_1}(\mathbf{x}_{l_1})^{1/a_{l_1}} + \cdots + n_{l_j}(\mathbf{x}_{l_j})^{1/a_{l_j}}. \tag{11.5}$$

Notice that intrinsic to the flag structure is an ordering of the subspaces $\mathbb{R}^{C_j}$, which must be taken into account in the labeling of variables, norms etc.

DEFINITION 11.1. A flag kernel associated to the flag $\mathcal{F}$ and the dilations $\lambda \cdot_{\boldsymbol{\alpha}} \mathbf{x}$ is a distribution $\mathcal{K}$ on $\mathbb{R}^N$ with the following size estimates and cancellation conditions:

(A) [Differential Inequalities] Away from the subspace where $\mathbf{x}_1 = 0$ the distribution $\mathcal{K}$ is given by integration against a smooth function K which satisfies

$$\left|\partial_{\mathbf{x}_1}^{\gamma_1} \cdots \partial_{\mathbf{x}_n}^{\gamma_n} K(\mathbf{x}_1, \ldots, \mathbf{x}_n)\right| \leq C_{\boldsymbol{\gamma}} \prod_{j=1}^{n} N_j'(\mathbf{x}_1, \ldots, \mathbf{x}_j)^{-a_j(Q_j + [\![\boldsymbol{\gamma}_j]\!])}$$

(B) [Cancellation Conditions] Let $L = \{l_1, \ldots, l_r\}$ and $M = \{m_1, \ldots, m_s\}$ be complementary subsets of $\{1, \ldots, n\}$, let $R = \{R_1, \ldots, R_s\}$ be positive real numbers, and let $\psi \in \mathcal{C}_0^\infty(\mathbb{R}^M)$ be a normalized bump function with support in the unit ball. Define a distribution $\mathcal{K}_{\psi,R}$ on $\mathbb{R}^{C_{l_1}} \oplus \cdots \oplus \mathbb{R}^{C_{l_r}}$ by setting $\langle \mathcal{K}_{\psi,R}, \varphi \rangle = \langle \mathcal{K}, \varphi \otimes \psi_R \rangle$ for every $\varphi \in \mathcal{C}_0^\infty(\mathbb{R}^L)$. Then $\mathcal{K}_{\psi,R}$ uniformly satisfies the analogue of the estimates in (a) on the space $\mathbb{R}^L$; *i.e.* away from the subspacee of $\mathbb{R}^L$ where $\mathbf{x}_{c_r} = 0$, the distribution $\mathcal{K}_{\psi,R}$ is given by integration against a smooth function $K_{\psi,R}$ and for every $\boldsymbol{\gamma} = (\gamma_{l_1}, \ldots, \gamma_{l_r}) \in \mathbb{N}^{C_{l_1}} \times \cdots \times \mathbb{N}^{C_{l_r}}$ there is a constant $C_{\boldsymbol{\gamma}}$ depending only on the constants in (a) and in particular independent of ψ and R so that

$$\left|\partial_{\mathbf{x}_{l_1}}^{\gamma_{l_1}} \cdots \partial_{\mathbf{x}_{l_r}}^{\gamma_{l_r}} K_{\psi,R}(\mathbf{x}_{l_1}, \ldots, \mathbf{x}_{l_r})\right| \leq C_{\boldsymbol{\gamma}} \prod_{j=1}^{r} N_{L,j}'(\mathbf{x}_{l_1}, \ldots, \mathbf{x}_{l_j})^{-a_{l_j}(Q_{l_j} + [\![\boldsymbol{\gamma}_{l_j}]\!])}.$$

As usual, if $s = n$, we require that $|\langle \mathcal{K}, \psi_R \rangle|$ is bounded independently of R.

The following is Theorem 2.3.9 in [**NRS01**].

THEOREM 11.2. *If $\mathcal{K}$ is a flag kernel adapted to the flag* (11.1) *and the family of dilations* (11.2), *the Fourier transform $m = \widehat{\mathcal{K}}$ is function smooth away from the subspace where $\boldsymbol{\xi}_n = 0$ which satisfies the differential inequalities*

$$\left|\partial_{\boldsymbol{\xi}_1}^{\gamma_1} \cdots \partial_{\boldsymbol{\xi}_m}^{\gamma_n} m(\boldsymbol{\xi}_1, \ldots, \boldsymbol{\xi}_n)\right| \leq C_{\boldsymbol{\gamma}} \prod_{j=1}^{n} \left(n_j(\boldsymbol{\xi}_j)^{1/a_j} + \cdots + n_n(\boldsymbol{\xi}_n)^{1/a_n}\right)^{-a_j[\![\boldsymbol{\gamma}_j]\!]}. \tag{11.6}$$

Conversely, any function satisfying (11.6) *is the Fourier transform of a flag kernel adapted to the flag* (11.1) *and dilations* (11.2).

11.2. Pairs of opposite flags

In Section 11.1, we chose a particular ordering of the subspaces $\{\mathbb{R}^{C_1}, \ldots, \mathbb{R}^{C_n}\}$ to define the flag $\mathcal{F}$ in (11.1) and a particular family of dilations on $\mathbb{R}^N$ in (11.2). We can consider the *opposite flag* obtained by reversing the indices $\{1, \ldots, n\}$ and choosing a different one-parameter family of dilations. Our objective is then to study *two-flag distributions* which are simultaneously flag kernels for the two different flags and two different homogeneities. Thus set

$$\begin{aligned} \mathcal{F}: &\quad (0) \subsetneq \mathbb{R}^{C_n} \subsetneq \mathbb{R}^{C_{n-1}} \oplus \mathbb{R}^{C_n} \subsetneq \cdots \subsetneq \mathbb{R}^{C_2} \oplus \cdots \oplus \mathbb{R}^{C_n} \subsetneq \mathbb{R}^N, \\ \mathcal{F}^\perp &\quad (0) \subsetneq \mathbb{R}^{C_1} \subsetneq \mathbb{R}^{C_1} \oplus \mathbb{R}^{C_2} \subsetneq \cdots \subsetneq \mathbb{R}^{C_1} \oplus \cdots \oplus \mathbb{R}^{C_{n-1}} \subsetneq \mathbb{R}^N. \end{aligned} \tag{11.7}$$

Let $\mathbf{a} = (a_1, \ldots, a_n)$ and $\mathbf{b} = (b_1, \ldots, b_n)$ be two n-tuples of positive real numbers. Define two one-parameter families of dilations

$$\begin{aligned} \lambda \cdot_{\mathbf{a}} \mathbf{x} &= (\lambda^{a_1} \cdot \mathbf{x}_1, \ldots, \lambda^{a_n} \cdot \mathbf{x}_n), \\ \lambda \cdot_{\mathbf{b}} \mathbf{x} &= (\lambda^{b_1} \cdot \mathbf{x}_1, \ldots, \lambda^{b_n} \cdot \mathbf{x}_n), \end{aligned} \tag{11.8}$$

and associate the first dilation with the flag $\mathcal{F}$ and the second with the flag $\mathcal{F}^\perp$. Along with satisfying the appropriate cancellation conditions, a two-flag distribution $\mathcal{K}$ then satisfies the following differential inequalities. Away from the set where $\mathbf{x}_1 = 0$ or where $\mathbf{x}_n = 0$, $\mathcal{K}$ is given by integration against a smooth function K and

$$\begin{aligned} &\big|\partial^\gamma K(\mathbf{x})\big| \\ &\leq C_\gamma \begin{cases} \prod_{j=1}^n \Big(n_1(\mathbf{x}_1)^{a_j/a_1} + \cdots + n_{j-1}(\mathbf{x}_{j-1})^{a_j/a_{j-1}} + n_j(\mathbf{x}_j)\Big)^{-(Q_j + [\![\gamma_j]\!])} \\ \prod_{j=1}^n \Big(n_j(\mathbf{x}_j) + n_{j+1}(\mathbf{x}_{j+1})^{b_j/b_{j+1}} + \cdots + n_n(\mathbf{x}_n)^{b_j/b_n}\Big)^{-(Q_j + [\![\gamma_j]\!])} \end{cases}. \end{aligned}$$

The corresponding Fourier transform $m = \widehat{\mathcal{K}}$ satisfies

$$\begin{aligned} &\big|\partial^\gamma m(\boldsymbol{\xi})\big| \\ &\leq C_\gamma \begin{cases} \prod_{j=1}^n \Big(n_j(\boldsymbol{\xi}_j) + n_{j+1}(\boldsymbol{\xi}_{j+1})^{a_j/a_{j+1}} + \cdots + n_n(\boldsymbol{\xi}_n)^{a_j/a_n}\Big)^{-[\![\gamma_j]\!]} \\ \prod_{j=1}^n \Big(n_1(\boldsymbol{\xi}_1)^{b_j/b_1} + \cdots + n_{j-1}(\boldsymbol{\xi}_{j-1})^{b_j/b_{j-1}} + n_j(\boldsymbol{\xi}_j)\Big)^{-[\![\gamma_j]\!]} \end{cases}. \end{aligned} \tag{11.9}$$

11.3. Two-step flags

We first consider the case of two-step flags, where $\mathbb{R}^N$ is written as a direct sum of two subspaces: $\mathbb{R}^N = \mathbb{R}^n \oplus \mathbb{R}^m$. As we will see, this situation is much simpler than the general case considered later in Section 11.4. Write elements of $\mathbb{R}^N$ as $(\mathbf{x}, \mathbf{y})$ with $\mathbf{x} \in \mathbb{R}^n$ and $\mathbf{y} \in \mathbb{R}^m$. For notational convenience we assume isotropic dilations on each component. There are two flags $\mathcal{F}$ and $\mathcal{F}^\perp$ adapted to this decomposition:

$$\mathcal{F}: \quad (0) \subsetneq \mathbb{R}^n \subsetneq \mathbb{R}^n \oplus \mathbb{R}^m = \mathbb{R}^N \quad \text{and} \quad \mathcal{F}^\perp: \quad (0) \subsetneq \mathbb{R}^m \subsetneq \mathbb{R}^n \oplus \mathbb{R}^m = \mathbb{R}^N.$$

Let $\mathbf{a} = (a_1, a_2)$ and $\mathbf{b} = (b_1, b_2)$ be two pairs of positive real numbers and put

$$\begin{aligned} \lambda \cdot_{\mathbf{a}} (\mathbf{x}, \mathbf{y}) &= (\lambda^{a_1}\mathbf{x}, \lambda^{a_2}\mathbf{y}), \\ \lambda \cdot_{\mathbf{b}} (\mathbf{x}, \mathbf{y}) &= (\lambda^{b_1}\mathbf{x}, \lambda^{b_2}\mathbf{y}). \end{aligned} \tag{11.10}$$

Associate the family of dilations $\lambda \cdot_a (\mathbf{x}, \mathbf{y})$ with the flag $\mathcal{F}$ and the family $\lambda \cdot_b (\mathbf{x}, \mathbf{y})$ with the flag $\mathcal{F}^\perp$. A distribution $\mathcal{K}$ on $\mathbb{R}^n \times \mathbb{R}^m$ is then a *two-flag kernel* relative to this data if it satisfies appropriate cancellation conditions and is given by integration against a smooth function K which satisfies the differential inequalities

$$\big|\partial_{\mathbf{x}}^{\boldsymbol{\alpha}} \partial_{\mathbf{y}}^{\boldsymbol{\beta}} K(\mathbf{x}, \mathbf{y})\big| \leq C_{\boldsymbol{\alpha},\boldsymbol{\beta}} \begin{cases} |\mathbf{x}|^{-n-|\boldsymbol{\alpha}|}\big(|\mathbf{x}|^{a_2/a_1} + |\mathbf{y}|\big)^{-m-|\boldsymbol{\beta}|} & \text{for } \mathbf{x} \neq 0 \text{ (flag } \mathcal{F}) \\ \big(|\mathbf{x}| + |\mathbf{y}|^{b_1/b_2}\big)^{-n-|\boldsymbol{\alpha}|} |\mathbf{y}|^{-m-|\boldsymbol{\beta}|} & \text{for } \mathbf{y} \neq 0 \text{ (flag } \mathcal{F}^\perp) \end{cases} \tag{11.11}$$

If $m(\boldsymbol{\xi},\boldsymbol{\eta}) = \widehat{\mathcal{K}}(\boldsymbol{\xi},\boldsymbol{\eta})$ is the Fourier transform, then
(11.12)

$$\left|\partial_{\xi}^{\boldsymbol{\alpha}}\partial_{\eta}^{\boldsymbol{\beta}}m(\boldsymbol{\xi},\boldsymbol{\eta})\right| \leq C_{\boldsymbol{\alpha},\boldsymbol{\beta}} \begin{cases} \big(|\boldsymbol{\xi}| + |\boldsymbol{\eta}|^{a_1/a_2}\big)^{-|\boldsymbol{\alpha}|}|\boldsymbol{\eta}|^{-|\boldsymbol{\beta}|} & \text{for } \boldsymbol{\eta} \neq 0 \text{ (flag } \mathcal{F}) \\ \\ |\boldsymbol{\xi}|^{-|\boldsymbol{\alpha}|}\big(|\boldsymbol{\xi}|^{b_2/b_1} + |\boldsymbol{\eta}|\big)^{-|\boldsymbol{\beta}|} & \text{for } \boldsymbol{\xi} \neq 0 \text{ (flag } \mathcal{F}^{\perp}) \end{cases}.$$

It follows that both K and m are smooth away from the origin.

PROPOSITION 11.3. Suppose that $\frac{a_1}{b_1} < \frac{a_2}{b_2}$, and let

$$N_1(\mathbf{x},\mathbf{y}) = |\mathbf{x}| + |\mathbf{y}|^{b_1/b_2} \qquad \widehat{N}_1(\boldsymbol{\xi},\boldsymbol{\eta}) = |\boldsymbol{\xi}| + |\boldsymbol{\eta}|^{a_1/a_2}$$
$$N_2(\mathbf{x},\mathbf{y}) = |\mathbf{x}|^{a_2/a_1} + |\mathbf{y}| \qquad \widehat{N}_2(\boldsymbol{\xi},\boldsymbol{\eta}) = |\boldsymbol{\xi}|^{b_2/b_1} + |\boldsymbol{\eta}|$$

be the norms associated with the matrix

$$\mathbf{E} = \begin{bmatrix} 1 & \frac{b_1}{b_2} \\ \frac{a_2}{a_1} & 1 \end{bmatrix} \tag{11.13}$$

satisfying the basic hypotheses.

If $\mathcal{K}$ is a two-flag kernel on $\mathbb{R}^n \times \mathbb{R}^m$ adapted to the dilations given in (11.10), then

(a) the corresponding function K is integrable at infinity;
(b) $\mathcal{K} \in \mathcal{P}(\mathbf{E})$ and $\widehat{\mathcal{K}} = m \in \mathcal{M}(\mathbf{E})$;
(c) we can write $\mathcal{K} = \mathcal{K}_0 + K_\infty$ where $K_\infty \in L^1(\mathbb{R}^N) \cap \mathcal{C}^\infty(\mathbb{R}^N)$, and $\mathcal{K}_0$ is a two-flag kernel supported in $\mathbb{B}(1)$.

Conversely, if $\mathcal{K}$ is in $\mathcal{P}(\mathbf{E})$, then $\mathcal{K}$ is a two-flag kernel (and $\widehat{\mathcal{K}} = m$ a two-flag multiplier) on $\mathbb{R}^n \times \mathbb{R}^m$ adapted to the dilations given in (11.10).

PROOF. We have already observed that m is smooth away from the origin, and it follows from equation (11.12) that we have the following estimates for pure derivatives of m:

$$\left|\partial_{\boldsymbol{\xi}}^{\boldsymbol{\alpha}}m(\boldsymbol{\xi},\boldsymbol{\eta})\right| \lesssim \big(|\boldsymbol{\xi}| + |\boldsymbol{\eta}|^{a_1/a_2}\big)^{-|\boldsymbol{\alpha}|},$$
$$\left|\partial_{\boldsymbol{\eta}}^{\boldsymbol{\beta}}m(\boldsymbol{\xi},\boldsymbol{\eta})\right| \lesssim \big(|\boldsymbol{\xi}|^{b_2/b_1} + |\boldsymbol{\eta}|\big)^{-|\boldsymbol{\beta}|}$$

Since the rank of $\mathbf{E}$ is greater than 1, it follows from Lemma 7.5 that K is smooth away from the origin and is integrable at infinity. This establishes (a).

Next let $\chi \in \mathcal{C}_0^\infty(\mathbb{R}^n \times \mathbb{R}^n)$ have support in the unit ball, with $\chi(\mathbf{x},\mathbf{y}) \equiv 1$ if $|\mathbf{x}| + |\mathbf{y}| \leq \frac{1}{2}$. Obviously $\mathcal{K}_0 = \chi\mathcal{K}$ satisfies the differential inequalities (11.11) and it easy to verify that it also satisfies the cancellations in Definition 11.1. By (a), $\mathcal{K}_\infty = (1-\chi)\mathcal{K} \in L^1(\mathbb{R}^n) \cap \mathcal{C}^\infty(\mathbb{R}^n)$. This establishes (c).

Being a two-flag kernel we know that K_0 satisfies (11.11). A case by case analysis shows that these estimates can be combined and improved.

(1) Suppose $|\mathbf{x}| \geq |\mathbf{y}|^{b_1/b_2}$. Then $|\mathbf{x}|^{-1} \lesssim \big(|\mathbf{x}|+|\mathbf{y}|^{b_1/b_2}\big)^{-1}$ and hence the first inequality shows that

$$|\partial_{\mathbf{x}}^{\boldsymbol{\alpha}}\partial_{\mathbf{y}}^{\boldsymbol{\beta}}K_0(\mathbf{x},\mathbf{y})| \lesssim \big(|\mathbf{x}| + |\mathbf{y}|^{b_1/b_2}\big)^{-n-|\boldsymbol{\alpha}|}\big(|\mathbf{x}|^{a_2/a_1} + |\mathbf{y}|\big)^{-m-|\boldsymbol{\beta}|}. \tag{11.14}$$

(2) Suppose $|\mathbf{y}| \geq |\mathbf{x}|^{a_2/a_1}$. Then $|\mathbf{y}|^{-1} \lesssim \big(|\mathbf{x}|^{a_2/a_1} + |\mathbf{y}|\big)^{-1}$, and the second inequality again gives the estimate in (11.14).

(3) Finally suppose $|\mathbf{x}| < |\mathbf{y}|^{b_1/b_2}$ and $|\mathbf{y}| < |\mathbf{x}|^{a_2/a_1}$. Here we must have

$$|\mathbf{x}| < |\mathbf{x}|^{\frac{a_2 b_1}{a_1 b_2}} \quad \text{and} \quad |\mathbf{y}| < |\mathbf{y}|^{\frac{a_2 b_1}{a_1 b_2}}$$

Since $a_2 b_1 > a_1 b_2$, this means that $|\mathbf{x}| > 1$ and $|\mathbf{y}| > 1$, and this is outside the support of K_0. It follows that for all $(\mathbf{x}, \mathbf{y}) \in \mathbb{R}^n \oplus \mathbb{R}^m$ we have the *improved* estimates

$$|\partial_{\mathbf{x}}^{\boldsymbol{\alpha}} \partial_{\mathbf{y}}^{\boldsymbol{\beta}} K(\mathbf{x}, \mathbf{y})| \lesssim \big(|\mathbf{x}| + |\mathbf{y}|^{\frac{b_1}{b_2}}\big)^{-n-|\alpha|} \big(|\mathbf{x}|^{\frac{a_2}{a_1}} + |\mathbf{y}|\big)^{-m-|\boldsymbol{\beta}|}.$$

This shows that $\mathcal{K}_0 \in \mathcal{P}_0(\mathbf{E})$ since the cancellations conditions for the flag kernels are the same as the cancellation conditions for $\mathcal{P}_0(\mathbf{E})$. This establishes (b) and completes the proof.

The last part of the statement is obvious. □

REMARKS 11.4.

(1) For a matrix $\mathbf{E}$ of the form (11.13), the condition $\frac{a_1}{b_1} \leq \frac{a_2}{b_2}$ is equivalent to the basic hypotheses. If the inequality is an equality, $\mathbf{E}$ is reducible and the class $\mathcal{P}(\mathbf{E})$ consists of Calderón-Zygmund kernels.

The arguments used in the proof of Proposition 11.3 can adapted to prove that, if the opposite inequality $\frac{a_1}{b_1} > \frac{a_2}{b_2}$ holds, then a kernel $\mathcal{K} \in \mathcal{P}(\mathbf{E})$ (defined by the same differential inequalities and cancellations as in Definition 2.2) is integrable near 0 instead than at infinity. This suggests that the essentially local nature of the kernels strongly depends on the basic hypotheses.

(2) One may observe that a 2×2 matrix satisfying the basic hypotheses can always be put in the form (11.13) for appropriate a_j, b_j with $\frac{a_1}{b_1} \leq \frac{a_2}{b_2}$. This means that, for every two-fold decomposition $\mathbb{R}^N = \mathbb{R}^{C_1} \oplus \mathbb{R}^{C_2}$ and every $\mathbf{E}$ satisfying the basic hypotheses, the class $\mathcal{P}(\mathbf{E})$ conicides either with a class of two-flag kernels or with a class of Calderón-Zygmund kernels, depending on its rank.

As we will see next, this is specific of two-fold decomposition. However, it follows from Lemma 5.3 that the same also holds for general decompositions, as long as the reduced rank of $\mathbf{E}$ is not greater than 2.

11.4. General two-flag kernels

In the case of the two-step flags, the improved estimates for the kernel followed easily from the two-flag estimates simply by considering possible cases. However, if the number of norms is greater than two, this argument no longer works. Observe that the differential inequalities for the kernel K give *no* information when $\mathbf{x}_1 = \mathbf{x}_n = 0$ while other $\mathbf{x}_j \neq 0$. In particular, it does not follow from the inequalities alone that the kernel is non-singular away from the origin. Nevertheless, we have the following result.

THEOREM 11.5. *Suppose that* $\frac{a_1}{b_1} \leq \frac{a_2}{b_2} \leq \cdots \leq \frac{a_n}{b_n}$ *with at least one strict inequality, and let* $\mathcal{K}$ *be a two-flag kernel for the flags* $\mathcal{F}$ *and* $\mathcal{F}^{\perp}$ *with homogeneities* $\mathbf{a}$ *and* $\mathbf{b}$ *respectively.*

(a) *The function* K *is integrable at infinity, and we can write* $\mathcal{K} = \mathcal{K}_0 + K_\infty$ *where* $K_\infty \in L^1(\mathbb{R}^N) \cap \mathcal{C}^\infty(\mathbb{R}^N)$*, and* $\mathcal{K}_0$ *is a two-flag kernel supported in* $\mathbb{B}(1)$.

(b) *The kernel $\mathcal{K}_0$ belongs to the class $\mathcal{P}_0(\mathbf{E})$ associated to the matrix*

$$\mathbf{E} = \begin{bmatrix} 1 & b_1/b_2 & b_1/b_3 & \cdots & b_1/b_{n-1} & b_1/b_n \\ a_2/a_1 & 1 & b_2/b_3 & \cdots & b_2/b_{n-1} & b_2/b_n \\ a_3/a_1 & a_3/a_2 & 1 & \cdots & b_3/b_{n-1} & b_3/b_n \\ \vdots & \vdots & \vdots & \ddots & \vdots & \vdots \\ a_{n-1}/a_1 & a_{n-1}/a_2 & a_{n-1}/a_3 & \cdots & 1 & b_{n-1}/b_n \\ a_n/a_1 & a_n/a_2 & a_n/a_3 & \cdots & a_n/a_{n-1} & 1 \end{bmatrix}. \tag{11.15}$$

(c) *Conversely, every kernel in $\mathcal{P}_0(\mathbf{E})$ is a two-flag kernel for the flags $\mathcal{F}$ and $\mathcal{F}^\perp$ with homogeneities $\mathbf{a}$ and $\mathbf{b}$ respectively.*

PROOF. If two ratios $a_j/b_j, a_k/b_k$ are equal, the product $e(j,k)e(k,j)$ is 1. By Proposition 5.5, the matrix $\mathbf{E}$ can be reduced to $\mathbf{E}^\flat$ of the same type. We may then assume that all inequalities are strict.

The first part of the proof proceeds as the proof of Proposition 11.3. Combining the two kinds of estimates for pure derivatives of the Fourier transform m, we see that m satisfies $\left|\partial_{\boldsymbol{\xi}_j}^{\gamma_j} m(\boldsymbol{\xi})\right| \lesssim \widehat{N}_j(\boldsymbol{\xi})^{-[\![\gamma_j]\!]}$ where

$$\begin{aligned}\widehat{N}_j(\boldsymbol{\xi}) = n_1(\boldsymbol{\xi}_1)^{b_j/b_1} + \cdots + n_{j-1}(\boldsymbol{\xi}_{j-1})^{b_j/b_{j-1}} + n_j(\boldsymbol{\xi}_j) + \\ + n_{j+1}(\boldsymbol{\xi}_{j+1})^{a_j/a_{j+1}} + \cdots + n_n(\boldsymbol{\xi}_n)^{a_j/a_n}.\end{aligned}$$

We want to apply Lemma 7.5, and so we need to check that the matrix $\mathbf{E}$ satisfies the basic hypothesis of (2.10) and has rank greater than 1. We proceed to check this.

We want to show that $e(j,k) \le e(j,l)e(l,k)$ where $e(j,k) = \begin{cases} b_j/b_k & \text{if } j \le k \\ a_j/a_k & \text{if } j \ge k \end{cases}.$

Thus we need to show

$$e(j,k) = \frac{b_j}{b_k} \le \begin{cases} \frac{a_j}{a_l}\frac{b_l}{b_k} = e(j,l)e(l,k) & \text{if } l \le j \le k, \\ \frac{b_j}{b_l}\frac{b_l}{b_k} = e(j,l)e(l,k) & \text{if } j \le l \le k, \\ \frac{b_j}{b_l}\frac{a_l}{a_k} = e(j,l)e(l,k) & \text{if } j \le k \le l, \end{cases} \qquad \text{if } j \le k,$$

$$e(j,k) = \frac{a_j}{a_k} \le \begin{cases} \frac{b_j}{b_l}\frac{a_l}{a_k} = e(j,l)e(l,k) & \text{if } l \ge j \ge k, \\ \frac{a_j}{a_l}\frac{a_l}{a_k} = e(j,l)e(l,k) & \text{if } j \ge l \ge k, \\ \frac{a_j}{a_l}\frac{b_l}{b_k} = e(j,l)e(l,k) & \text{if } j \ge k \ge l, \end{cases} \qquad \text{if } j \ge k.$$

However it is easy to check that all of these inequalities follow from the assumption in the Theorem. Thus $\mathbf{E}$ satisfies the basic hypotheses. Also one can check that

$$\det(\mathbf{E}) = \prod_{j=2}^{n}\left(1 - \frac{a_j}{b_j}\frac{b_{j-1}}{a_{j-1}}\right) \ne 0$$

so in fact the matrix $\mathbf{E}$ has rank n. Thus it follows from Lemma 7.5 that K is integrable at infinity.

The proof of part (b) will use the following result, which shows that estimates for pure derivatives leads to estimates for mixed derivatives. In order to keep our focus on the proof of Theorem 11.5, we defer the proof to Section 11.5.

LEMMA 11.6. *Let $m \in \mathcal{C}^\infty(\mathbb{R}^N \setminus \{0\})$. Suppose that for every $\gamma_j \in \mathbb{N}^{C_j}$ there is a constant $C_{\gamma_j} > 0$ so that*

$$\left|\partial_{\boldsymbol{\xi}_j}^{\gamma_j} m(\boldsymbol{\xi})\right| \leq C_{\gamma_j} \widehat{N}_j(\boldsymbol{\xi})^{-[\![\gamma_j]\!]}.$$

Then for every $\gamma = (\gamma_1, \dots, \gamma_n) \in \mathbb{N}^{C_1} \times \cdots \times \mathbb{N}^{C_n}$ there is a constant $C_\gamma > 0$ so that if $\boldsymbol{\xi} \in \mathbb{B}(1)^c$,

$$\left|\partial_{\boldsymbol{\xi}_1}^{\gamma_1} \cdots \partial_{\boldsymbol{\xi}_n}^{\gamma_n} m(\boldsymbol{\xi})\right| \leq C_\gamma \prod_{j=1}^{n} \widehat{N}_j(\boldsymbol{\xi})^{-[\![\gamma_j]\!]}.$$

We now turn to the proof of Theorem 11.5 part (b), which is more involved than in the situation considered in Section 11.3. We start by decomposing the Fourier transform m_0 of the kernel $\mathcal{K}_0$ by means of a smooth cutoff function supported in $B(2)$ and equal to 1 on $B(1)$. We obtain $m_0 = m' + m''$, where m' is in $\mathcal{C}_0(\mathbb{R}^N)$, while m'' is supported in $B(1)^c$ and satisfies the inequalities (11.9). Denoting by $\mathcal{K}', \mathcal{K}''$ the inverse Fourier transforms of m', m'' respectively, we then have $\mathcal{K}_0 = \mathcal{K}' + \mathcal{K}''$.

We consider m'' first. Since it satisfies the inequalities (11.9), the two types of estimates can be combined together to give

$$\left|\partial_{\boldsymbol{\xi}_j}^{\gamma_j} m''(\boldsymbol{\xi})\right| \lesssim \widehat{N}_j(\boldsymbol{\xi})^{-[\![\gamma_j]\!]}, \tag{11.16}$$

whenever we differentiate only in the $\boldsymbol{\xi}_j$-variables (*pure* derivatives). Applying Lemma 11.6 and recalling that m'' is supported in $B(1)^c$, we obtain inequalities for all derivatives of m'', namely

$$\left|\partial_{\boldsymbol{\xi}_1}^{\gamma_1} \cdots \partial_{\boldsymbol{\xi}_n}^{\gamma_n} m''(\boldsymbol{\xi})\right| \lesssim \prod_{j=1}^{n} \left(1 + \widehat{N}_j(\boldsymbol{\xi})\right)^{-[\![\gamma_j]\!]}, \tag{11.17}$$

which tells us that $m'' \in \mathcal{M}_\infty(\mathbf{E})$.

By Theorem 4.1 it follows that $\mathcal{K}'' \in \mathcal{P}_0(\mathbf{E})$. Using a cutoff function in the $\mathbf{x}$-variables, we could then split $\mathcal{K}''$ as $\mathcal{K}_0 + \mathcal{K}'''$, with $\mathcal{K}_0 \in \mathcal{P}_0(\mathbf{E})$ and $\mathcal{K}''' \in \mathcal{S}(\mathbb{R}^N)$. We next observe that $\mathcal{K}' \in C^\infty(\mathbb{R}^n)$ because m' has compact support. It follows that $\mathcal{K}$ coincides with a smooth function away from the origin. We can then apply Theorem 7.5 and conclude that $\mathcal{K}$ is integrable at infinity. Hence $\mathcal{K}' \in C^\infty(\mathbb{R}^N) \cap L^1(\mathbb{R}^N)$ and the same holds for $\mathcal{K}' + \mathcal{K}''' = \mathcal{K}_\infty$. Finally, part (c) is an obvious consequence of (b). □

REMARK 11.7.

As mentioned in Remark 11.4, for a general matrix $\mathbf{E}$ with reduced rank larger than 2, it is not true that the class $\mathcal{P}(\mathbf{E})$ coincides with a class of two-flag kernels. In fact, such a coincidence occurs if and only if the equality $e(j,\ell) = e(j,k)e(k,\ell)$ holds whenever the triple j, k, ℓ is ordered ($j < k < l$ or vice-versa).

The following statement is a direct consequence of Theorem 9.2.

COROLLARY 11.8. Let $\mathcal{K}, \mathcal{L}$ be two compactly supported two-flag kernels as in Theorem 11.5, for the same pair of flags and the same homogeneities. Then the convolution $\mathcal{K} * \mathcal{L}$ is also a two-flag kernel of the same type.

11.5. Proof of Lemma 11.6

In this section we complete the proof of Theorem 11.5 by proving Lemma 11.6; *i.e.* we show that if $m \in \mathcal{C}^\infty(\mathbb{R}^N)$ satisfies the estimates for the class of multipliers $\mathcal{M}(\mathbf{E})$ for *pure* derivatives, then the estimates for mixed derivatives follow automatically. We will use the following result, which is easily proved by estimating the decay of the Fourier transform of f.

PROPOSITION 11.9. Let $f \in \mathcal{C}_0^\infty(\mathbb{R}^N)$ be supported in the unit ball. Suppose that

$$\left|\partial_{\boldsymbol{\eta}_j}^{\gamma_j} f(\boldsymbol{\eta})\right| \leq C_{\gamma_j}$$

for each j and each $\gamma_j \in \mathbb{N}^{C_j}$. Then for every $\gamma = (\gamma_1, \ldots, \gamma_n) \in \mathbb{N}^{C_1} \times \cdots \times \mathbb{N}^{C_n}$ we have

$$\left|\partial_{\boldsymbol{\eta}_1}^{\gamma_1} \cdots \partial_{\boldsymbol{\eta}_n}^{\gamma_n} f(\boldsymbol{\eta})\right| \leq C'_\gamma$$

where the constants $\{C'_\gamma\}$ depend only on the constants $\{C_{\gamma_j}\}$.

Next let $S = \{(I_1, k_1); \ldots; (I_s, k_s)\} \in \mathcal{S}(n)$. For $\boldsymbol{\xi} \in \mathbb{B}(1)^c \cap \widehat{E}_S^A$, we give a quantitative estimate of the size of a ball about $\boldsymbol{\xi}$ that we want contained in some larger region $\widehat{E}_S^{A'}$. Recall from Definition 3.12 and equation (3.5) that we defined a dilation structure $\lambda \widehat{\cdot}_S \boldsymbol{\xi}_{I_r} = \{\lambda^{1/e(k_r,j)} \cdot \boldsymbol{\xi}_j : j \in I_r\}$ and a corresponding smooth homogeneous norm $\widehat{n}_{S,r}(\boldsymbol{\xi}_{I_r}) \approx \sum_{j \in I_r} n_j(\boldsymbol{\xi}_j)^{e(k_r,j)}$ on $\mathbb{R}^{I_r}$. Let $C \geq 1$ be a constant such that, for all $1 \leq r \leq s$, we have

$$\widehat{n}_{S,r}(\boldsymbol{\xi}_{I_r} + \boldsymbol{\eta}_{I_r}) \leq C\left[\widehat{n}_{S,r}(\boldsymbol{\xi}_{I_r}) + \widehat{n}_{S,r}(\boldsymbol{\eta}_{I_r})\right]. \tag{11.18}$$

PROPOSITION 11.10. Let $S = \{(I_1, k_1); \ldots; (I_s, k_s)\} \in \mathcal{S}(n)$ and let C be the constant from equation (11.18). Let $\boldsymbol{\xi} = (\boldsymbol{\xi}_1, \ldots, \boldsymbol{\xi}_n) \in \mathbb{B}(1)^c \cap \widehat{E}_S^A$. If $\epsilon < (2C)^{-1}$, then the set

$$\Big\{\boldsymbol{\eta} \in \mathbb{R}^N : \widehat{n}_{S,r}(\boldsymbol{\xi}_{I_r} - \boldsymbol{\eta}_{I_r}) = \sum_{j \in I_r} n_j(\boldsymbol{\eta}_j - \boldsymbol{\xi}_j)^{e(k_r,j)} < \epsilon\, n_{k_r}(\boldsymbol{\xi}_{k_r}),\ 1 \leq r \leq s\Big\}$$

is contained in $\widehat{E}_S^{A'}$ where A' is a constant that depends only on A and C.

PROOF. Let $\boldsymbol{\xi} \in \widehat{E}_S^A$. Recall from part (b) of Lemma 3.9 that

$$\begin{aligned} n_{k_r}(\boldsymbol{\xi}_{k_r}) &> A^{-1} n_j(\boldsymbol{\xi}_j)^{e(k_r,j)} && \forall j \in I_r, j \neq k_r, \quad \text{and} \\ n_{k_r}(\boldsymbol{\xi}_{k_r}) &> A^{-1} n_{k_t}(\boldsymbol{\xi}_{k_t})^{1/\tau_S(k_r,k_t)} && \forall t \in \{1, \ldots, s\},\ t \neq r. \end{aligned}$$

Conversely, if $\boldsymbol{\eta} \in \mathbb{B}(1)^c$ satisfies

$$\begin{aligned} n_{k_r}(\boldsymbol{\eta}_{k_r}) &> A^{-1} n_j(\boldsymbol{\eta}_j)^{e(k_r,j)} && \forall j \in I_r, j \neq k_r, \quad \text{and} \\ n_{k_r}(\boldsymbol{\eta}_{k_r}) &> A^{-1} n_{k_t}(\boldsymbol{\eta}_{k_t})^{1/\tau_S(k_r,k_t)} && \forall t \in \{1, \ldots, s\},\ t \neq r. \end{aligned}$$

then $\boldsymbol{\eta} \in E_S^{A^\tau}$ where τ is a constant that depends only on the coefficients $\{e(j,k)\}$.

If $n_j(\boldsymbol{\eta}_j - \boldsymbol{\xi}_j)^{e(k_r,j)} < \epsilon n_{k_r}(\boldsymbol{\xi}_{k_r})$ for $1 \leq r \leq s$ and all $j \in I_R$, and if $C\epsilon < \frac{1}{2}$, then $n_{k_r}(\boldsymbol{\xi}_{k_r}) \leq C\big[n_{k_r}(\boldsymbol{\xi}_{k_r} - \boldsymbol{\eta}_{k_r}) + n_{k_r}(\boldsymbol{\eta}_{k_r})\big] \leq C\big[\epsilon n_{k_r}(\boldsymbol{\xi}_{k_r}) + n_{k_r}(\boldsymbol{\eta}_{k_r})\big] < \frac{1}{2} n_{k_r}(\boldsymbol{\xi}_{k_r}) + C n_{k_r}(\boldsymbol{\eta}_{k_r})$ so that for $1 \leq r \leq s$,

$$n_{k_r}(\boldsymbol{\xi}_{k_r}) < 2C n_{k_r}(\boldsymbol{\eta}_{k_r}). \tag{11.19}$$

Next, if $j \in I_r$,

$$n_j(\boldsymbol{\eta}_j) \leq C\big[n_j(\boldsymbol{\eta}_j - \boldsymbol{\xi}_j) + n_j(\boldsymbol{\xi}_j)\big] < C\Big[\big(\epsilon n_k(\boldsymbol{\xi}_{k_r})\big)^{1/e(k_r,j)} + \big(A n_{k_r}(\boldsymbol{\xi}_{k_r})\big)^{1/e(k_r,j)}\Big],$$

and it follows from (11.19) that

$$n_j(\boldsymbol{\eta}_j) \le C'\, n_{k_r}(\boldsymbol{\eta}_{k_r})^{1/e(k_r,j)} \tag{11.20}$$

where C' depends on C, ϵ, and A. Finally, if $1 \le t \le s$ with $t \ne r$, it follows from (11.20) that $n_{k_t}(\boldsymbol{\eta}_{k_t}) \le C' n_{k_t}(\boldsymbol{\xi}_{k_t})^{1/e(k_t,k_t)}$, and so

$$n_{k_t}(\boldsymbol{\eta}_{k_t}) \le C'\big(A n_{k_r}(\boldsymbol{\xi}_{k_r})\big)^{\tau_S(k_r,k_t)} \le C''\, n_{k_r}(\boldsymbol{\eta}_{k_r})^{\tau_S(k_r,k_t)} \tag{11.21}$$

where C'' depends on C, ϵ, and A. It now follows from the inequalities in (11.20) and (11.21) that $\boldsymbol{\eta} \in \widehat{E}_S^{A'}$, and this completes the proof. □

We now prove Lemma 11.6.

Let $\boldsymbol{\xi} \in \mathbb{B}(1)^c$. Then if $A > 1$ there exists $S = \big((I_1, k_1); \ldots; (I_S, k_s)\big) \in \mathcal{S}(n)$, so that $\boldsymbol{\xi} \in \widehat{E}_S^A$. We have $\widehat{N}_j(\boldsymbol{\xi}) \approx n_{k_r}(\boldsymbol{\xi}_{k_r})^{1/e(k_r,j)}$ for every $r \in \{1, \ldots, s\}$ and every $j \in I_r$. (Here the implied constants depend on A.) Thus to establish the Lemma we need to show that for every $\boldsymbol{\gamma} = (\gamma_1, \ldots, \gamma_n) \in \mathbb{N}^{C_1} \times \cdots \times \mathbb{N}^{C_n}$ there is a constant $C_{\boldsymbol{\gamma}} > 0$ so that

$$\big|\partial_{\boldsymbol{\xi}_1}^{\gamma_1} \cdots \partial_{\boldsymbol{\xi}_n}^{\gamma_n} m(\boldsymbol{\xi})\big| \le C_{\boldsymbol{\gamma}} \prod_{r=1}^{s} n_{k_r}(\boldsymbol{\xi}_{k_r})^{-\sum_{j\in I_r} [\![\gamma_j]\!]/e(k_r,j)}.$$

According to Proposition 11.10, there exists $\epsilon > 0$ and a constant A' depending only on A so that

$$\Big\{\boldsymbol{\eta} \in \mathbb{R}^N : n_j(\boldsymbol{\eta}_j - \boldsymbol{\xi}_j)^{e(k_r,j)} < \epsilon\, n_{k_r}(\boldsymbol{\xi}_{k_r}) \text{ for } 1 \le r \le s \text{ and } j \in I_r\Big\} \subset \widehat{E}_S^{A'}.$$

For $1 \le r \le s$ let $\varphi_r \in \mathcal{C}_0^\infty(\mathbb{R}^{I_r})$ be supported where $n_j(\boldsymbol{\eta}_j) \le 1$ for all $j \in I_r$, with $\widehat{n}_{S,r}(\boldsymbol{\eta}_{I_r}) \equiv 1$ if $n_j(\boldsymbol{\eta}_j) \le \frac{1}{2}$ for all $j \in I_r$. Put

$$f(\boldsymbol{\eta}) = \Big(\prod_{r=1}^{s} \varphi_r(\boldsymbol{\eta}_{I_r})\Big)\, m\Big(\boldsymbol{\xi}_{I_1} + [\epsilon\, n_{k_1}(\boldsymbol{\xi}_{k_1})]\hat{\cdot}_S\, \boldsymbol{\eta}_{I_1}, \ldots, \boldsymbol{\xi}_{I_S} + [\epsilon\, n_{k_s}(\boldsymbol{\xi}_{k_s})]\hat{\cdot}_S\, \boldsymbol{\eta}_{I_s}\Big).$$

Here, the function m is evaluated at the point whose j^{th}-coordinate is

$$\mathbf{v}_j = \boldsymbol{\xi}_j + [\epsilon n_{k_r}(\boldsymbol{\xi}_{k_r})]^{1/e(k_r,j)} \cdot \boldsymbol{\eta}_j$$

when $j \in I_r$. Clearly $f \in \mathcal{C}^\infty(\mathbb{R}^N)$ and is supported where $n_j(\boldsymbol{\eta}_j) \le 1$ for all $1 \le j \le n$. Moreover

$$f(\boldsymbol{\eta}) = m\Big(\boldsymbol{\xi}_{I_1} + [\epsilon\, n_{k_1}(\boldsymbol{\xi}_{k_1})]\hat{\cdot}_S\, \boldsymbol{\eta}_{I_1}, \ldots, \boldsymbol{\xi}_{I_S} + [\epsilon\, n_{k_s}(\boldsymbol{\xi}_{k_s})]\hat{\cdot}_S\, \boldsymbol{\eta}_{I_s}\Big)$$

if $n_j(\boldsymbol{\eta}_j) \le \frac{1}{2}$ for all j. If $f(\boldsymbol{\eta}) \ne 0$, so that $n_j(\boldsymbol{\eta}_j) \le 1$, then

$$\begin{aligned} n_j(\mathbf{v}_j - \boldsymbol{\xi}_j) &= n_j\Big(n_{k_r}([c\,\boldsymbol{\xi}_{k_r})]^{1/e(k_r,j)} \cdot \boldsymbol{\eta}_j\Big) \\ &= [\epsilon\, n_{k_r}(\boldsymbol{\xi}_{k_r})]^{1/e(k_r,j)}\, n_j(\boldsymbol{\eta}_j) \le [\epsilon\, n_{k_r}(\boldsymbol{\xi}_{k_r})]^{1/e(k_r,j)}, \end{aligned}$$

and it follows that m is evaluated at a point of $\widehat{E}_S^{A'}$.

Consider 'pure' derivatives $\partial_{\boldsymbol{\eta}_j}^{\gamma_j} f$. Some will fall on $\big(\prod_{r=1}^{s} \varphi_r(\boldsymbol{\eta}_{I_r})\big)$ and the rest fall on the term involving m. Derivatives of $\big(\prod_{r=1}^{s} \varphi_r(\boldsymbol{\eta}_{I_r})\big)$ are bounded by constants independent of m. On the other hand, if $\boldsymbol{\gamma}_j \in \mathbb{N}^{(I_r)}$, using the chain rule, the hypothesis about pure derivatives of m, and the fact that we are evaluating m at points in $\widehat{E}_S^{A'}$, we get

$$\begin{aligned}
&\partial_{\boldsymbol{\eta}_j}^{\gamma_j}\Big[m\Big(\boldsymbol{\xi}_{I_1}+[\epsilon\, n_{k_1}(\boldsymbol{\xi}_{k_1})]\hat{\cdot}_S\,\boldsymbol{\eta}_{I_1},\ldots,\boldsymbol{\xi}_{I_S}+[\epsilon\, n_{k_s}(\boldsymbol{\xi}_{k_s})]\hat{\cdot}_S\,\boldsymbol{\eta}_{I_s}\Big)\Big]\\
&\qquad=\Big[\partial_{\boldsymbol{\eta}_j}^{\gamma_j}m\Big]\Big(\boldsymbol{\xi}_{I_1}+[\epsilon\, n_{k_1}(\boldsymbol{\xi}_{k_1})]\hat{\cdot}_S\,\boldsymbol{\eta}_{I_1},\ldots,\boldsymbol{\xi}_{I_S}+[\epsilon\, n_{k_s}(\boldsymbol{\xi}_{k_s})]\hat{\cdot}_S\,\boldsymbol{\eta}_{I_s}\Big)\times\\
&\qquad\qquad\prod_{j\in I_r}\big[\epsilon\, n_{k_r}(\boldsymbol{\xi}_{k_r})\big]^{[\![\gamma_j]\!]/e(k_r,j)}\\
&\qquad\leq C_{\gamma_j}\prod_{j\in I_r}n_{k_r}(\boldsymbol{\xi}_{k_r})^{-[\![\gamma_j]\!]/e(k_r,j)}\prod_{j\in I_r}\big[\epsilon\, n_{k_r}(\boldsymbol{\xi}_{k_r})\big]^{[\![\gamma_j]\!]/e(k_r,j)}\leq C_{\gamma_j}.
\end{aligned}$$

Thus $f \in \mathbb{C}^\infty(\mathbb{R}^N)$ is supported in the unit ball, and pure derivatives are bounded. It follows from Proposition 11.9 that all derivatives are bounded by constants depending only on the $\{C_{\gamma_j}\}$. But for $n_j(\boldsymbol{\eta}_j) \leq \frac{1}{2}$, we have

$$\partial_{\boldsymbol{\eta}_1}^{\gamma_1}\cdots\partial_{\boldsymbol{\eta}_n}^{\gamma_n}f(\boldsymbol{\eta})=[\partial_{\boldsymbol{\eta}_1}^{\gamma_1}\cdots\partial_{\boldsymbol{\eta}_n}^{\gamma_n}m](\mathbf{v}_1,\ldots,\mathbf{v}_n)\prod_{r=1}^{s}n_{k_r}(\boldsymbol{\xi}_{k_r})^{\sum_{j\in I_r}[\![\gamma_j]\!]/e(k_r,j)}.$$

where $\mathbf{v}_j = \boldsymbol{\xi}_j + [\epsilon n_{k_r}(\boldsymbol{\xi}_{k_r})]^{1/e(k_r,j)} \cdot \boldsymbol{\eta}_j$. When $\boldsymbol{\eta} = 0$, the uniform bounds on f give the desired estimates for the derivatives of m, and this completes the proof. □

CHAPTER 12

Extended kernels and operators

Operators of the form $T_{\mathcal{K}} f = f * \mathcal{K}$ with $\mathcal{K} \in \mathcal{P}_0(\mathbf{E})$ are "constant coefficient"operators. In this chapter we consider a more general class of operators which have "variable coefficients". Identify the space $\mathbb{R}^N$ with a nilpotent Lie group $G \cong \mathbb{R}^N$ as in Section 8, and consider kernels of the form $K(\mathbf{x}, \mathbf{z})$, with $(\mathbf{x}, \mathbf{z}) \in G \times G$, so that for each $\mathbf{x} \in G$, $K(\mathbf{x}, \cdot) \in \mathcal{P}(\mathbf{E})$, with smooth dependence on $\mathbf{x}$. More precisely, assume that for each norm $|\cdot|_M$ defined in (2.13) and each β,

$$\sup_{\mathbf{x}} |X^\beta \mathcal{K}(\mathbf{x}, \cdot)|_M < \infty \,, \tag{12.1}$$

where X^β is a monomial in the left-invariant vector fields of G acting on the $\mathbf{x}$-variable. We also assume (unless $G = \mathbb{R}^N$ as an additive group) the hypotheses of Theorems 8.9 and 9.2 to guarantee that proper kernels define bounded convolution on L^p, $1 < p < \infty$ and form a class closed under convolution.

Kernels of this kind will be said to belong to the "extended class", to distinguish them from the "proper" kernels in $\mathcal{P}_0(\mathbf{E})$ which are independent of $\mathbf{x}$. Operators corresponding to these extended kernels are given formally by

$$Tf(\mathbf{x}) = \int_G K(\mathbf{x}, \mathbf{y}^{-1}\mathbf{x}) f(\mathbf{y}) d\mathbf{y}; \tag{12.2}$$

that is, if $f \in \mathcal{S}(\mathbb{R}^N)$, $Tf(\mathbf{x}) = \langle K(\mathbf{x}, \cdot), f_{\mathbf{x}} \rangle$, with $f_{\mathbf{x}}(\mathbf{y}) = f(\mathbf{x}\mathbf{y}^{-1})$.

Note that the extended class of operators is left-invariant in the following sense. If $\mathbf{a} \in G$ set $L_{\mathbf{a}} f(\mathbf{x}) = f(\mathbf{a}^{-1}\mathbf{x})$. Then $L_{\mathbf{a}}^{-1} T L_{\mathbf{a}}$ is an operator again of the form (12.2), with $K(\mathbf{x}, \mathbf{z})$ replaced by $K(\mathbf{a}\mathbf{x}, \mathbf{z})$. Observe that the semi-norms (12.1) of $K(\mathbf{x}, \mathbf{z})$ and $K(\mathbf{a}\mathbf{x}, \mathbf{z})$ are the same.

Our main results concerning this extended class of kernels and operators are as follows.

THEOREM 12.1. *Each operator T given by (12.2) is bounded on $L^p(\mathbb{R}^N)$, for $1 < p < \infty$.*

THEOREM 12.2. *The class of these operators form an algebra under addition and composition of operators.*

12.1. The L^p boundedness of the operators

Recall that if $\delta_r : G \to G$ is the one-parameter group of automorphic dilations of the group G, the coordinates on $\mathbb{R}^N$ are chosen so that $\delta_r(x_1, \ldots, x_N) = (r^{d_1} x_1, \ldots, r^{d_N} x_N)$ with $d_1 \le d_2 \le \cdots \le d_N$. Let $|\cdot|_G$ be a smooth homogeneous norm, so that $|\delta_r \mathbf{x}|_G = r\,|\mathbf{x}|_G$. The homogeneous dimension of G is $Q = \sum_{j=1}^N d_j$, and $d(\mathbf{x}, \mathbf{y}) = |\mathbf{y}^{-1}\mathbf{x}|_G$ is a left-invariant quasi-distance. Let

$B_r(\mathbf{x}) = \{\mathbf{y} \in G : d(\mathbf{x}, \mathbf{y}) < r\}$. Then $|B_r(\mathbf{x})| = |B_r(0)| = r^Q |B_1(0)|$. We will write $B_r(0) = B_r$.

In proving Theorem 12.1, we make a preliminary reduction to kernels $\mathcal{K}(\mathbf{x}, \mathbf{z})$ that are supported when $\mathbf{z} \in B_1$. In fact let η be a C^∞ function supported in B_1 which equals 1 when $\mathbf{z} \in B_{1/2}$. Then writing $\mathcal{K}(\mathbf{x}, \mathbf{z}) = \mathcal{K}_0(\mathbf{x}, \mathbf{z}) + \mathcal{K}_\infty(\mathbf{x}, \mathbf{z}) = \mathcal{K}(\mathbf{x}, \mathbf{z})\eta(\mathbf{z}) + \mathcal{K}(\mathbf{x}, \mathbf{z})(1 - \eta(\mathbf{z}))$, we see that $K_\infty(\mathbf{x}, \mathbf{z})$ is bounded and rapidly decreasing as $|\mathbf{z}| \to \infty$. Now the operator T_∞, that is (12.2) with $\mathcal{K}_\infty$ in place at $\mathcal{K}$, can be written as $T_\infty(f)(\mathbf{x}) = \int K_\infty(\mathbf{x}, \mathbf{y}) f(\mathbf{x}\mathbf{y}^{-1}) d\mathbf{y}$, and hence, by Minkowski's inequality, is bounded on L^p, for all $1 \le p \le \infty$.

Thus suppose $\mathcal{K}(\mathbf{x}, \mathbf{z})$ is supported when $\mathbf{z} \in B_1$. We make a further assumption that $\mathcal{K}(\mathbf{x}, \mathbf{z})$ is supported when $\mathbf{x} \in B_\lambda$, where λ is so small so that $B_{2\lambda}$ is strictly contained in the torus $\mathbb{T} = \{\mathbf{x} : |x_j| < \frac{1}{2},\ j = 1, \ldots, N\}$. (Later we will lift this hypothesis.) In this case we can extend $\mathcal{K}$ outside of $\mathbb{T}$ by periodicity, and expand $\mathcal{K}(\mathbf{x}, \mathbf{z})$ as Fourier series in the $\mathbf{x}$-variable,

$$\mathcal{K}(\mathbf{x}, \mathbf{z}) = \sum_{\mathbf{k} \in \mathbb{Z}^N} e^{2\pi i \mathbf{k} \cdot \mathbf{x}} \mathcal{H}_{\mathbf{k}}(\mathbf{z}) ,$$

with

$$\mathcal{H}_{\mathbf{k}}(\mathbf{z}) = \frac{1}{(2\pi)^N} \int_{\mathbb{T}} e^{-2\pi i \mathbf{k} \cdot \mathbf{x}} K(\mathbf{x}, \mathbf{z}) d\mathbf{x} . \tag{12.3}$$

Each kernel $\mathcal{H}_{\mathbf{k}}(\mathbf{z})$ can be written as $\alpha_{\mathbf{k}} \mathcal{K}_{\mathbf{k}}(\mathbf{z})$ in such a way that the $\alpha_{\mathbf{k}}$ are rapidly decreasing coefficients and the $\mathcal{K}_{\mathbf{k}}$ are proper kernels, supported for $\mathbf{z} \in B_1$, whose semi-norms $|\mathcal{K}_{\mathbf{k}}|_M$ are uniformly bounded in $\mathbf{k}$ for every M. This can be done via the following diagonal argument. Integrating by parts in (12.3), we obtain that, for every integer q,

$$|\mathcal{H}_{\mathbf{k}}|_M \le \frac{C_q}{|\mathbf{k}|^q} |\mathcal{K}|_{q,M} ,$$

where $|\mathcal{K}|_{q,M}$ denotes the supremum of the semi-norms implied by (12.1) over $|\beta| \le q$. Then, given q, we inductively choose increasing integers m_q such that

$$\frac{C_q}{m_q^q} |\mathcal{K}|_{q,q} < \frac{1}{m_q^{q/2}} .$$

For $m_q \le |\mathbf{k}| < m_{q+1}$, we set

$$\alpha_{\mathbf{k}} = \frac{C_q}{|\mathbf{k}|^q} |\mathcal{K}|_{q,q} .$$

Then $\alpha_{\mathbf{k}} < |\mathbf{k}|^{-q/2}$, showing that the $\alpha_{\mathbf{k}}$ are rapidly decreasing, and the kernels $\mathcal{K}_{\mathbf{k}} = \alpha_{\mathbf{k}}^{-1} \mathcal{H}_{\mathbf{k}}$ are uniformly bounded in every M-norm (we remark that the M-norms are increasing in M). Upon relabeling the series and with a slight abuse of notation, this gives

$$\mathcal{K}(\mathbf{x}, \mathbf{z}) = \sum_{n=1}^{\infty} a_n(\mathbf{x}) \mathcal{K}_n(\mathbf{z}) . \tag{12.4}$$

Here $a_n(\mathbf{x}) = \eta(\mathbf{x}) \alpha_{\mathbf{k}} e^{2\pi i \mathbf{k} \cdot \mathbf{x}}$, $\mathcal{K}_n(\mathbf{z}) = \mathcal{K}_{\mathbf{k}}(\mathbf{z})$ (for a $\mathbf{k}$ that depends on n); also η is a suitable C^∞ cut-off function supported in $\mathbf{x} \in B_\lambda$. With this choice, the $a_n(\mathbf{x})$ are a rapidly decreasing sequence of C^∞ functions, each supported in B_λ. Clearly now the operator T whose kernel is given by (12.4) is bounded on L^p, in view of Theorem 8.9.

Now, with the above λ fixed, we will drop the restriction that $\mathcal{K}(\mathbf{x},\mathbf{z})$ is supported for $\mathbf{x} \in B_\lambda$, by a further decomposition which writes $\mathbb{R}^N$ as a union of balls $B_\lambda(\mathbf{x}_m) = \mathbf{x}_m \cdot B_\lambda$, with these balls chosen so they are "almost"disjoint. For this we need the following lemma.

LEMMA 12.3. *Given $\lambda > 0$, there is a collection $\{B_\lambda(\mathbf{x}_m)\}$ so that*

(1) *$\{B_{\lambda/2}(\mathbf{x}_m)\}$ covers $\mathbb{R}^N$;*
(2) *for every $\mu > 0$, the collection $\{B_\mu(\mathbf{x}_m)\}$ has the bounded intersection property: there exists an integer $M = M_\mu$ so that no point $\mathbf{x} \in \mathbb{R}^M$ is contained in more than M of the $\{B_\mu(\mathbf{x}_m)\}$.*

PROOF. Fix a constant c_1 (which is small with λ, and which will be chosen momentarily) and choose the centers $\mathbf{x}_m$ so that the balls $\{B_{c_1}(\mathbf{x}_m)\}$ are disjoint, and so that this is a maximal family with respect to that property. Now if $\bigcup\limits_m B_{\lambda/2}(\mathbf{x}_m)$ does not cover $\mathbb{R}^N$, there exists an $\mathbf{x}_0$, so that $d(\mathbf{x}_0, \mathbf{x}_m) \geq \lambda/2$ for all m. We then claim that $B_{c_1}(\mathbf{x}_0)$ must be disjoint from all the $B_{c_1}(\mathbf{x}_m)$, violating the maximality. In fact if this disjointness fails, then there is an $\mathbf{x}' \in B_{c_1}(\mathbf{x}_0) \cap B_{c_1}(\mathbf{x}_m)$ for some m. This means $d(\mathbf{x}', \mathbf{x}_0) < c_1$ and $d(\mathbf{x}', \mathbf{x}_m) < c_1$, which by the quasi-triangle inequality $d(\mathbf{x}_0, \mathbf{x}_m) \leq c(d(\mathbf{x}', \mathbf{x}_0) + d(\mathbf{x}', \mathbf{x}_m))$ gives a contradiction with $d(\mathbf{x}_0, \mathbf{x}_m) \geq \lambda/2$, when $\lambda/2 \geq 2cc_1$. Thus if we choose $c_1 = \lambda/4c$ we get conclusion (1).

To prove the second conclusion we may choose $\mu \geq c_1$. Suppose $\mathbf{x}'$ and $\mathbf{x}''$ belong to $B_\mu(\mathbf{x}_m)$. Then $d(\mathbf{x}', \mathbf{x}'') \leq 2c\mu$, so $\mathbf{x}'' \in B_{2c\mu}(\mathbf{x}')$, and hence $B_\mu(\mathbf{x}_m) \subset B_{2c\mu}(\mathbf{x}')$. So if $\mathbf{x}'$ belongs to M of the $B_\mu(\mathbf{x}_m)$, then $B_\mu(\mathbf{x}_m) \subset B_{2c\mu}(\mathbf{x}')$ for each of these m. Since then $B_{c_1}(\mathbf{x}_m) \subset B_{2c\mu}(\mathbf{x}')$ and the $B_{c_1}(\mathbf{x}_m)$ are disjoint, we have

$$\sum m\left(B_{c_1}(\mathbf{x}_m)\right) \leq m\left(B_{2c\mu}(\mathbf{x}')\right),$$

where the sum is taken over those m. By homogeneity we get $Mc_1^Q \leq (2c\mu)^Q$, and conclusion (2) is established with $M = (2c/c_1)^Q$. □

We now return to the proof of Theorem 12.1. Let us fix a positive C^∞ function φ that equals 1 in $B_{\lambda/2}$ and vanishes outside B_λ. Let $\varphi_m(\mathbf{x}) = \varphi\left(\mathbf{x}_m^{-1} \cdot \mathbf{x}\right)$, with the $\mathbf{x}_m$ as given in Lemma (12.3). Then because of conclusions (1) and (2) of the lemma, $1 \leq \sum \varphi_m(\mathbf{x}) \leq M$. Set $\psi_m(\mathbf{x}) = \varphi_m(\mathbf{x}) / \sum \varphi_m(\mathbf{x})$. Since $\{X^\alpha \varphi_m\}$ are bounded uniformly in m, the same is true of $\{\psi_m\}$. Notice that the $\{\psi_m\}$ give a partition of unity, with ψ_m supported in $B_\lambda(\mathbf{x}_m)$.

We let $\mathcal{K}_m(\mathbf{x},\mathbf{z}) = \psi_m(\mathbf{x})\mathcal{K}(\mathbf{x},\mathbf{z})$ and $T_m(f)(\mathbf{x}) = \int \mathcal{K}_m(\mathbf{x}, \mathbf{y}^{-1}\mathbf{x}) f(\mathbf{y}) d\mathbf{y}$. Then $T = \sum T_m$. Observe that T_m is of the form $L_{\mathbf{x}_m} T' L_{\mathbf{x}_m}^{-1}$, where $L_\mathbf{x}$ is left-translation by $\mathbf{x}$ and T' is an operator for which we already know its L^p boundedness by Theorem 8.9. Hence

$$||T_m(f)||_{L^p} \leq A_p ||f||_{L^p}, \tag{12.5}$$

uniformly in m, for $1 < p < \infty$. Now because $T = \sum T_m$, the lemma (with $\mu = \lambda$) gives us

$$\int |T|^p d\mathbf{x} \leq M_\lambda \sum_m \int |T_m(f)|^p d\mathbf{x}$$

However by (12.5),

$$\int |T_m(f)|^p \, d\mathbf{x} \leq A_p^p \int_{B_\mu(\mathbf{x}_m)} |f|^p d\mathbf{x}.$$

This is because $\mathbf{x} \in B_\lambda(\mathbf{x}_m)$ and $\mathbf{y}^{-1}\mathbf{x} \in B_1$ imply $\mathbf{y} \in B_\mu(\mathbf{x}_m)$ here with $\mu = c(\lambda+1)$, in view of the fact that $d(\mathbf{y},\mathbf{x}_m) \leq c(d(\mathbf{x},\mathbf{x}_m)+d(\mathbf{y},\mathbf{x}))$, and the support properties of $K_m(\mathbf{x},\mathbf{z})$. Again by conclusion (2) of Lemma 12.3,

$$\sum_m \int_{B_\mu(\mathbf{x}_m)} |f|^p d\mathbf{x} \leq M_\mu \int |f|^p d\mathbf{x},$$

giving the conclusion $||T(f)||_{L^p} \leq A'_p ||f||_{L^p}$ and so proving Theorem 12.1.

12.2. The algebra of operators

Proceeding to the proof of Theorem 12.2, we first summarize and slightly rephrase the essential idea of the previous section. Whenever $\mathcal{K}(\mathbf{x},\mathbf{z})$ is a kernel of the extended class that is supported for $|\mathbf{z}| \leq 1$, then we can write it as

$$\mathcal{K}(\mathbf{x},\mathbf{z}) = \sum_{n,m} a_{n,m}(\mathbf{x})\mathcal{K}_{n,m}(\mathbf{z}) \ , \tag{12.6}$$

where

(a) the $\mathcal{K}_{n,m}$ are proper kernels, each supported for $\mathbf{z}$ in the unit ball, with M-norms that are uniform in n and m;
(b) the $a_{n,m}(\mathbf{x})$ are C^∞ functions, each supported in $\mathbf{x} \in B_\lambda(\mathbf{x}_m)$;
(c) for every α, $\sup_{\mathbf{x}} |X^\alpha a_{n,m}(\mathbf{x})|$ is rapidly decreasing as $n \to \infty$, uniformly in m.

To prove Theorem 12.2 we study the composition of two operators T and S given by

$$T(f)(\mathbf{x}) = \int \mathcal{K}(\mathbf{x},\mathbf{y}^{-1}\mathbf{x})f(\mathbf{y})d\mathbf{y}, \quad S(f)(\mathbf{x}) = \int \mathcal{L}(\mathbf{x},\mathbf{y}^{-1}\mathbf{x})f(\mathbf{y})d\mathbf{y}$$

where $\mathcal{K}$ and $\mathcal{L}$ are both kernels of the extended class. We consider first their essential parts, that is, we restrict ourselves to $K(\mathbf{x},\mathbf{z})$ and $L(\mathbf{x},\mathbf{z})$ that are supported when $\mathbf{z}$ is in the unit ball B_1. Now by (12.6), $T = \sum_{n,m} T_{n,m}$ and $S = \sum_{n',m'} S_{n',m'}$ where

$$T_{n,m}(f)(\mathbf{x}) = a_{n,m}(\mathbf{x})(f * \mathcal{K}_{n,m})(\mathbf{x}), \quad S_{n',m'}(f)(\mathbf{x}) = b_{n',m'}(\mathbf{x})(f * \mathcal{L}_{n',m'})(\mathbf{x}).$$

and where $\{\mathcal{K}_{n,m}\}$ and $\{\mathcal{L}_{n',m'}\}$ are uniform families of proper kernels, each supported in $|\mathbf{z}| \leq 1$.

We next observe that, for each m, $T_{n,m} \circ S_{n',m'} = 0$ unless m' belongs to a subset of boundedly many m' (depending on m). In fact, by the support properties of $a_{n,m}$, $b_{n',m'}$, and the kernels $\mathcal{K}_{n,m}$, the product $T_{n,m} \cdot S_{n',m'}$ is non-vanishing only if the quasi-distance between $B_\lambda(\mathbf{x}_m)$ and $B_\lambda(\mathbf{x}_{m'})$ does not exceed 1. Thus by conclusion (2) of Lemma 12.3, this only happens for at most M_μ of the m', where $\mu = c^2(2\lambda+1)$.

From this, and the rapid decrease of $a_{n,m}$ and $b_{n',m'}$ as $n \to \infty$ and $n' \to \infty$, we see that the analysis of the composition can be reduced to the case when both T and S are each one summand. So we write $T = a \circ U$, $S = b \circ V$, where $U(f) = f * \mathcal{K}$ and $V(f) = f * \mathcal{L}$, with both $\mathcal{K}$ and $\mathcal{L}$ proper kernels supported for $|\mathbf{z}| \leq 1$. Now

$$T \circ S = a \circ U \circ b \circ V = a \circ b \circ U \circ V + a \circ [b, U] \circ V.$$

The term $a\circ b\circ U\circ V$ is of the right form because it corresponds to the kernel $a(\mathbf{x})b(\mathbf{x})(\mathcal{L}*\mathcal{K})(\mathbf{z})$, and $\mathcal{L}*\mathcal{K}$ is a proper kernel by Theorem 9.2. On the other hand,

$$[b,U](f)(\mathbf{x}) = \int \big(b(\mathbf{x})-b(\mathbf{y})\big)\mathcal{K}(\mathbf{y}^{-1}\mathbf{x})f(\mathbf{y})\,d\mathbf{y}\,. \tag{12.7}$$

Write $b(\mathbf{x})-b(\mathbf{y}) = c(\mathbf{x},\mathbf{z})$, with $\mathbf{z}=\mathbf{y}^{-1}\mathbf{x}$; thus $c(\mathbf{x},\mathbf{z}) = b(\mathbf{x})-b(\mathbf{x}\mathbf{z}^{-1})$. Recall that $c(\mathbf{x},\mathbf{z})$ can be restricted to $\mathbf{z}\in B_1$, (the support of K), and so can be taken to have compact $\mathbf{z}$–support. Also, $c(\mathbf{x},\mathbf{z})$ is jointly C^∞, and $\sup_{\mathbf{x}}\big|X_{\mathbf{x}}^{\alpha}\partial_{\mathbf{z}}^{\beta}c(\mathbf{x},\mathbf{z})\big|<\infty$, uniformly in $\mathbf{z}$.

At this stage we invoke the following lemma.

LEMMA 12.4. *Let $\mathcal{K}\in\mathcal{P}_0(\mathbf{E})$ and $c\in\mathcal{C}_0^\infty(\mathbb{R}^N)$. Then $c\mathcal{K}\in\mathcal{P}_0(\mathbf{E})$.*

PROOF. We use the Fourier transform characterization of Theorems 4.1 and 4.2: if $m=\widehat{\mathcal{K}}$, then $m\in\mathcal{M}_\infty(\mathbf{E})$ and

$$\widehat{c\mathcal{K}}(\boldsymbol{\xi}) = m*\widehat{c}\,(\boldsymbol{\xi}) = \int_{\mathbb{R}^N} m(\boldsymbol{\xi}-\boldsymbol{\eta})\widehat{c}(\boldsymbol{\eta})d\boldsymbol{\eta}.$$

We have $\widehat{N}_j(\boldsymbol{\xi}+\boldsymbol{\eta}) \le C\big[\widehat{N}_j(\boldsymbol{\xi})+\widehat{N}_j(\boldsymbol{\eta})\big]$, and if $|\boldsymbol{\xi}|$ is the Euclidean norm of $\boldsymbol{\xi}$, there are positive real numbers a and b so that $|\boldsymbol{\xi}|^a\le\widehat{N}_j(\boldsymbol{\xi})\le|\boldsymbol{\xi}|^b$ for $1\le j\le n$ and all $|\boldsymbol{\xi}|\ge 1$. Let $0<\delta<ab^{-1}$ and write

$$\begin{aligned}\partial^\gamma(m*\widehat{c}\,) &= \int_{|\boldsymbol{\eta}|\le|\boldsymbol{\xi}|^\delta}\partial_{\boldsymbol{\xi}}^{\gamma}m(\boldsymbol{\xi}-\boldsymbol{\eta})\,\widehat{c}\,(\boldsymbol{\eta})d\boldsymbol{\eta} + \int_{|\boldsymbol{\eta}|\ge|\boldsymbol{\xi}|^\delta}\partial_{\boldsymbol{\xi}}^{\gamma}m(\boldsymbol{\xi}-\boldsymbol{\eta})\widehat{c}(\boldsymbol{\eta})d\boldsymbol{\eta}\\ &= \mathrm{I}+\mathrm{II}.\end{aligned}$$

If $|\boldsymbol{\eta}|\le|\boldsymbol{\xi}|^\delta$ then $\widehat{N}_j(\boldsymbol{\xi}-\boldsymbol{\eta})\approx\widehat{N}_j(\boldsymbol{\xi})$ for all $1\le j\le n$ and all sufficiently large $|\boldsymbol{\xi}|$, and hence

$$|I| \le \int_{|\boldsymbol{\eta}|<|\boldsymbol{\xi}|^\delta}\prod_{j=1}^{n}(1+\widehat{N}_j(\boldsymbol{\xi}-\boldsymbol{\eta}))^{-[\![\gamma_j]\!]}\,|\widehat{c}(\boldsymbol{\eta})|\,d\boldsymbol{\eta} \lesssim \prod_{j=1}^{n}\big(1+\widehat{N}_j(\boldsymbol{\xi})\big)^{-[\![\gamma_j]\!]}.$$

On the other hand, because of the rapid decrease of $\widehat{c}$,

$$|II| \le c_M|\boldsymbol{\xi}|^{-M} \le c_M\prod_{j=1}^{n}\big(1+\widehat{N}_j(\boldsymbol{\xi})\big)^{-[\![\alpha_j]\!]}$$

if $M\ge\sum_{j=1}^{n}[\![\alpha_j]\!]$. As a result, $c\mathcal{K}\in\mathcal{P}_0(\mathbf{E})$. □

We now apply Lemma 12.4 to the kernel $c(\mathbf{x},\mathbf{z})\mathcal{K}(\mathbf{z})$ for each fixed $\mathbf{x}$. This shows that $c(\mathbf{x},\mathbf{z})\mathcal{K}(\mathbf{z})$ is an extended kernel. An examination of the argument guarantees that every semi-norm (12.1) of this kernel is controlled by only finitely many of the semi-norms $\sup_{\mathbf{x},\mathbf{z}}\big|\partial_{\mathbf{x}}^{\alpha}\partial_{\mathbf{z}}^{\beta}c(\mathbf{x},\mathbf{z})\big|$ of c and the M-norms of $\mathcal{K}$.

Coming to the operator $[b,U]\circ V$, we see that

$$([b,U]\circ V)(f)(\mathbf{x}) = \int\mathcal{M}(\mathbf{x},\mathbf{y}^{-1}\mathbf{x})f(\mathbf{y})d\mathbf{y},$$

where $\mathcal{M}(\mathbf{x},\,\cdot\,)=\mathcal{L}*\widetilde{K}(\mathbf{x},\cdot)$, where $V(f)=f*\mathcal{L}$, and where $\widetilde{K}(\mathbf{x},\mathbf{z})=c(\mathbf{x},\mathbf{z})K(\mathbf{z})$.

Then $\mathcal{M}(\mathbf{x},\mathbf{z})$ is an extended kernel, in view of Theorem 9.2; moreover any semi-norm (12.1) of $\mathcal{M}$ is controlled by finitely many semi-norms of $b,\mathcal{K}$ and $\mathcal{L}$. Finally, the rapid decrease of the $a_{n,m}$, as $n\to\infty$ (and the corresponding decrease of the $b_{n',m'}$ as $n' \to \infty$) then assures that $T \circ S$ is an operator of the extended class under the assumption to both $\mathcal{K}(\mathbf{x}, \mathbf{z})$ and $\mathcal{L}(\mathbf{x}, \mathbf{z})$ are supported when $|\mathbf{z}| \leq 1$.

The case when either one or both $\mathcal{K}(\mathbf{x}, \mathbf{z})$ and $\mathcal{L}(\mathbf{x}, \mathbf{z})$ are supported for $\mathbf{z}$ away from the origin is a simpler version of the above argument. In effect, when $\mathcal{K}$ is supported when $|\mathbf{z}| \leq 1$, and $\mathcal{L}$ is supported when $|\mathbf{z}| \geq c > 0$, then we need only use the facts that a convolution of a Schwartz function with a distribution of compact support is a Schwartz function, or that the convolution of two Schwartz functions is again a Schwartz function. This allows us to finish the proof of Theorem 12.2.

CHAPTER 13

The role of pseudo-differential operators

In this chapter we study the interplay of pseudo-differential operators with the operators of the extended class treated in Section 12. While more general cases can also be treated, for simplicity we restrict ourselves to the situation where the homogeneous group underlying our Euclidean space $\mathbb{R}^N$ is a stratified group, and where the matrix $\mathbf{E}$ arises from the two-flag example in Theorem 11.5 where $a_1 = \cdots = a_n = 1$ and $b_j = \frac{1}{j}$. Thus we take

$$\mathbf{E} = \begin{bmatrix} 1 & 2 & 3 & \cdots & n-1 & n \\ 1 & 1 & 3/2 & \cdots & (n-1)/2 & n/2 \\ 1 & 1 & 1 & \cdots & (n-1)/3 & n/3 \\ \vdots & \vdots & \vdots & \ddots & \vdots & \vdots \\ 1 & 1 & 1 & \cdots & 1 & n/(n-1) \\ 1 & 1 & 1 & \cdots & 1 & 1 \end{bmatrix}.$$

In this case of a stratified group we have the double identification of $\mathbb{R}^N = \mathbb{R}^{C_1} \oplus \cdots \oplus \mathbb{R}^{C_n}$ with the underlying group and also (via exponential coordinates) with its Lie algebra $\mathfrak{g}$. We also assume that the subspace $\mathbb{R}^{C_k} = \mathfrak{h}_k$ is the subspace homogeneous of degree k under the automorphic dilations δ_r. The fact that $\mathfrak{g}$ is stratified guarantees that $\mathfrak{h}_{k+1} = [\mathfrak{h}_k, \mathfrak{h}_1]$. Suppose the cardinality of C_k is c_k. If $\mathbf{x} = (\mathbf{x}_1, \ldots, \mathbf{x}_n) \in \mathbb{R}^N$ with $\mathbf{x}_k \in \mathbb{R}^{C_k}$, write $\mathbf{x}_k = (x_{k,1}, \ldots, x_{k,c_k})$. The automorphic dilations on $\mathfrak{h}_k$ are given by $\delta_\lambda \mathbf{x}_k = (\lambda^k x_{k,1}, \ldots, \lambda^k x_{k,c_k})$, the homogeneous dimension of $\mathfrak{h}_k$ is $k\,c_k$, and for $\boldsymbol{\alpha}_k \in \mathbb{N}^{c_k}$, we have $[\![\boldsymbol{\alpha}_k]\!] = k\,|\boldsymbol{\alpha}_k|$, where $|\boldsymbol{\alpha}_k|$ is the standard length of the multi-index.

We take the homogeneous norm on $\mathfrak{h}_k$ to be

$$n_k(\mathbf{t}_k) = |\mathbf{t}_k|^{\frac{1}{k}} = \Big[\sum_{j=1}^{C_k} |t_{k,j}|^2\Big]^{1/2k},$$

where $|\cdot|$ denotes the Euclidean norm. Then

$$\begin{aligned} N_j(\mathbf{x}) &= |\mathbf{x}_1| + |\mathbf{x}_2|^{\frac{1}{2}} + \cdots + |\mathbf{x}_{j-1}|^{\frac{1}{j-1}} + |\mathbf{x}_j|^{\frac{1}{j}} + |\mathbf{x}_{j+1}|^{\frac{1}{j}} + \cdots + |\mathbf{x}_n|^{\frac{1}{j}} \\ \widehat{N}_j(\boldsymbol{\xi}) &= |\boldsymbol{\xi}_1|^{\frac{1}{j}} + |\boldsymbol{\xi}_2|^{\frac{1}{j}} + \cdots |\boldsymbol{\xi}_{j-1}|^{\frac{1}{j}} + |\boldsymbol{\xi}_j|^{\frac{1}{j}} + |\boldsymbol{\xi}_{j+1}|^{\frac{1}{j+1}} + \cdots + |\boldsymbol{\xi}_n|^{\frac{1}{n}} \end{aligned} \tag{13.1}$$

Notice that $N_n = \widehat{N}_0$ is a homogeneous norm for the automorphic dilations δ_λ, and N_1 is equal to the Euclidean norm.

We will consider standard pseudo-differential operators of order 0. By this we mean operators of the form

$$f \mapsto P(f)(\mathbf{x}) = \int_{\mathbb{R}^N} a(\mathbf{x}, \boldsymbol{\xi})\, \widehat{f}\,(\boldsymbol{\xi}) e^{2\pi i \mathbf{x}\cdot\boldsymbol{\xi}} d\boldsymbol{\xi}, \tag{13.2}$$

defined for Schwartz functions f, where the symbol $a(\mathbf{x},\boldsymbol{\xi})$ is assumed to be of compact support in $\mathbf{x}$ and satisfy the differential inequalities

$$\left|\partial_{\mathbf{x}}^{\beta}\partial_{\boldsymbol{\xi}}^{\alpha}a(\mathbf{x},\boldsymbol{\xi})\right| \leq A_{\alpha,\beta}(1+|\boldsymbol{\xi}|)^{-|\alpha|}\,. \tag{13.3}$$

In what follows the errors introduced by commutators will be *smoothing operators* in the following sense. For $1<p<\infty$, the space L_1^p consists of all f in L^p so that Xf (taken in the weak sense) belongs to $L^p(G)$, for each left-invariant vector field X of the generating sub-space $\mathfrak{h}_1$ of $\mathfrak{g}$. We define a norm on L_1^p by setting

$$||f||_{L_1^p} = ||f||_{L^p} + \sum_{j=1}^{g} ||X_j(f)||_{L^p}, \tag{13.4}$$

where $\{X_1,\dots,X_g\}$ a basis for the subspace $\mathfrak{h}_1$. We shall also see below that an $f\in L^p$ belongs to L_1^p if and only if

$$||f(\mathbf{xh})-f(\mathbf{x})||_{L^p} = O\big(N_n(\mathbf{h})\big), \quad \text{as } \mathbf{h}\to 0. \tag{13.5}$$

An operator will be said to be *smoothing* if it maps L^p to L_1^p, for all $1<p<\infty$. The following theorem will be proved in the next sections.

THEOREM 13.1. *Suppose P is a standard pseudo-differential operator of order 0. Then P belongs to the extended class treated in Section* 12. *Moreover if T is any operator of this class, then the commutator $[T,P]$ is a smoothing operator.*

13.1. The isotropic extended kernels

For the proof of Theorem 13.1 it is necessary to consider a special sub-class $\mathcal{CZ}_0\subset\mathcal{P}_0(\mathbf{E})$, (the isotropic Calderón-Zygmund kernels on $\mathbb{R}^N$ which agree with a Schwartz function at infinity) and a corresponding sub-class of the "extended" kernels. A distribution $\mathcal{K}\in\mathcal{CZ}_0$ is assumed to satisfy the stronger inequalities

$$|\partial^{\gamma}K(\mathbf{z})| \leq A_{\alpha}|\mathbf{z}|^{-N-|\gamma|}\big(1+|\mathbf{z}|\big)^{-M}\,, \tag{13.6}$$

for every γ and M. Moreover these kernels are required to satisfy only the one type of cancellation property which arises when $n=1$. It follows from Theorems 4.1 and 4.2 (for the case $n=1$) that this is equivalent to the condition that the Fourier transform $m=\widehat{\mathcal{K}}$ satisfies $|\partial^{\gamma}m(\boldsymbol{\xi})|\leq A'_{\alpha}(1+|\boldsymbol{\xi}|)^{-|\gamma|}$. But then it follows from equation (13.1) that $m\in\mathcal{M}_{\infty}(\mathbf{E})$, and hence $\mathcal{CZ}_0\subset\mathcal{P}_0(\mathbf{E})$. We call elements of $\mathcal{CZ}_0$ *isotropic proper kernels.*

Together with $\mathcal{CZ}_0$ we consider the extended kernels $\mathcal{K}(\mathbf{x},\mathbf{z})$ such that for $\mathbf{x}$ fixed, $\mathcal{K}(\mathbf{x},\,\cdot\,)\in\mathcal{CZ}_0$. More precisely, we shall assume that these have compact support in $\mathbf{x}$ and all $\mathbf{x}$-derivatives $\partial_{\mathbf{x}}^{\alpha}\mathcal{K}(\mathbf{x},\,\cdot\,)$ are isotropic proper kernels uniformly in $\mathbf{x}$. Notice that, having assumed compact support in $\mathbf{x}$, this condition can be equivalently formulated with the $\mathbf{x}$-derivatives replaced by left-invariant, or right-invariant, vector fields. We will refer to these kernels as *isotropic extended kernels.*

The key fact we need is the following.

PROPOSITION 13.2. Suppose P is a pseudo-differential operator (13.2) of order 0, with symbol $a(\mathbf{x},\boldsymbol{\xi})$ having compact support in $\mathbf{x}$. Then P can be represented in the following three alternative forms: for any Schwartz function f:

(a) $P(f)(\mathbf{x}) = \int\limits_{\mathbb{R}^N}\mathcal{K}(\mathbf{x},\mathbf{x}-\mathbf{y})f(\mathbf{y})d\mathbf{y};$

(b) $P(f)(\mathbf{x}) = \int\limits_{\mathbb{R}^N}\mathcal{K}^L(\mathbf{x},\mathbf{y}^{-1}\mathbf{x})f(\mathbf{y})d\mathbf{y};$

(c) $P(f)(\mathbf{x}) = \int\limits_{\mathbb{R}^N} \mathcal{K}^R(\mathbf{x}, \mathbf{x}\mathbf{y}^{-1}) f(\mathbf{y}) d\mathbf{y}$.

Here $\mathcal{K}$, $\mathcal{K}^L$, and $\mathcal{K}^R$ are each isotropic extended kernels.

The realization (a) is well-known; see for instance [**Ste93**], Sections 4 and 7.4 of Chapter 6. In this case $\widehat{\mathcal{K}}(\mathbf{x}, \cdot) = a(\mathbf{x}, \cdot)$ (where the Fourier transform acts on the second variable), and reduces to the fact that a multiplier $m(\boldsymbol{\xi})$ that satisfies the differential inequalities $\left|\partial_{\boldsymbol{\xi}}^{\alpha} m(\boldsymbol{\xi})\right| \leq A_\alpha (1 + |\boldsymbol{\xi}|)^{-|\alpha|}$ arises as the Fourier transform of an isotropic proper kernel, and so this is covered by Theorems 4.1 and 4.2. The other two representations need further analysis.

First we assert that

$$\mathbf{y}^{-1}\mathbf{x} = L_{\mathbf{x},\mathbf{y}}(\mathbf{x} - \mathbf{y}), \tag{13.7}$$

where for each $(\mathbf{x}, \mathbf{y}) \in G \times G$, $L_{\mathbf{x},\mathbf{y}}$ is a linear transformation that depends polynomially on $\mathbf{x}$ and $\mathbf{y}$. In fact, by the homogeneity of our group and using the Campbell-Hausdorff formula we can write for each $1 \leq k \leq n$ and each $1 \leq j \leq c_k$,

$$\left(\mathbf{y}^{-1}\mathbf{x}\right)_{k,j} = x_{k,j} - y_{k,j} + \sum_{\ell < k} \sum_{j=1}^{c_\ell} p_k^{\ell,j}(\mathbf{x}, \mathbf{y}) \cdot (x_{\ell,j} - y_{\ell,j})$$

with $p_k^{\ell,j}$ polynomials in $\mathbf{x}$ and $\mathbf{y}$ jointly homogeneous at degree $k - \ell$ (see for example Section 6.6 in [**NRSW12**], or Chapter 1 of [**FS82**]). This gives (13.7) with $L_{\mathbf{x},\mathbf{y}}$ represented by a block-triangular matrix which in particular implies that $\det(L_{\mathbf{x},\mathbf{y}}) = 1$, for all $\mathbf{x}$ and $\mathbf{y}$. Now let $M_{\mathbf{x},\mathbf{y}}$ denote the inverse of $L_{\mathbf{x},\mathbf{y}}$. Then $\mathbf{x} - \mathbf{y} = M_{\mathbf{x},\mathbf{y}}(\mathbf{y}^{-1}\mathbf{x})$ and $M_{\mathbf{x},\mathbf{y}}$ is again polynomially dependent on $\mathbf{x}$ and $\mathbf{y}$.

Define a (non-linear) mapping $\Phi = \Phi_{\mathbf{x}} : \mathbb{R}^N \to \mathbb{R}^N$ by setting $\Phi_{\mathbf{x}}(\mathbf{z}) = M_{\mathbf{x},\mathbf{x}\mathbf{z}^{-1}}(\mathbf{z})$. Given $\mathbf{x}$ and $\mathbf{w}$,

$$\Phi_{\mathbf{x}}\left((\mathbf{x} - \mathbf{w})^{-1}\mathbf{x}\right) = M_{\mathbf{x},\mathbf{x}\mathbf{x}^{-1}(\mathbf{x}-\mathbf{w})}\left((\mathbf{x} - \mathbf{w})^{-1}\mathbf{x}\right) = \mathbf{x} - (\mathbf{x} - \mathbf{w}) = \mathbf{w}$$

so $\Phi_{\mathbf{x}}$ maps $\mathbb{R}^N$ onto $\mathbb{R}^N$ and the inverse of Φ_x is given by $\Phi_{\mathbf{x}}^{-1}(\mathbf{w}) = (\mathbf{x} - \mathbf{w})^{-1}\mathbf{x} = L_{\mathbf{x},\mathbf{x}-\mathbf{w}}(\mathbf{w})$. Thus both $\Phi_{\mathbf{x}}$ and $\Phi_{\mathbf{x}}^{-1}$ are polynomial mappings of $\mathbb{R}^N$ to itself (and hence diffeomorphisms) that each depend polynomially on $\mathbf{x}$. In our applications of these mappings below, the range of $\mathbf{x}$ is restricted to a compact set, in view of the support hypothesis on the symbol $a(\mathbf{x}, \boldsymbol{\xi})$ and the resulting fact that $\mathcal{K}(\mathbf{x}, \mathbf{z})$ has compact-support in $\mathbf{x}$.

We set down the following two simple properties of the mappings $\Phi_{\mathbf{x}}$:

1) $|\Phi_{\mathbf{x}}(\mathbf{z})| \approx |\mathbf{z}|$, for $|\mathbf{z}| \leq 1$;

2) $|\mathbf{z}|^{\frac{1}{n}} \lesssim |\Phi_{\mathbf{x}}(\mathbf{z})| \lesssim |\mathbf{z}|^n$, for $|\mathbf{z}| \geq 1$.

The first assertion is a consequence of the fact that Φ is a diffeomorphism that fixes the origin. For the second, we see first that $|\Phi_{\mathbf{x}}(\mathbf{z})| \leq c|\mathbf{z}|^n$, $|\mathbf{z}| \geq 1$ because of the polynomial dependence (of degree n) of the linear mappings $M_{\mathbf{x},\mathbf{y}}$. Since the same inequality holds for the inverse $\Phi_{\mathbf{x}}^{-1}$ we get the reverse inequality $|\mathbf{z}|^{\frac{1}{n}} \leq c|\Phi_{\mathbf{x}}(\mathbf{z})|$, again for large $|\mathbf{z}|$.

Our proof of the second representation (b) in Proposition 13.2 starts with knowing the first representation and defining $\mathcal{K}^L(\mathbf{x}, \mathbf{z}) = \mathcal{K}(\mathbf{x}, \Phi_{\mathbf{x}}(\mathbf{z}))$. Then, at least formally, we have

$$\int \mathcal{K}^L(\mathbf{x}, \mathbf{y}^{-1}\mathbf{x}) f(\mathbf{y}) d\mathbf{y} = \int \mathcal{K}(\mathbf{x}, \mathbf{x} - \mathbf{y}) f(\mathbf{y}) d\mathbf{y} = P(f)(\mathbf{x}). \tag{13.8}$$

In order to justify this it will suffice to see that $\mathcal{K}^L(\mathbf{x},\mathbf{z})$ is, for each $\mathbf{x}$, an isotropic proper kernel, and (in both variables) an isotropic extended kernel. For this we make the following observation.

LEMMA 13.3. *Let* $\mathcal{K}(\mathbf{z}) \in \mathcal{CZ}_0$. *Suppose* Φ *is a* C^∞ *diffeomorphism of* $\mathbb{R}^N$ *fixing the origin, and assume that* $|\Phi(\mathbf{z})| \gtrsim |\mathbf{z}|^\delta$, *for some* $\delta > 0$ *and all* $|\mathbf{z}| \geq 1$. *Then* $\mathcal{K}\circ\Phi \in \mathcal{CZ}_0$ *where the distribution* $\mathcal{K}\circ\Phi$ *is defined by* $\langle \mathcal{K}\circ\Phi,\ \varphi\rangle = \langle \mathcal{K}, \varphi(\Phi^{-1})J_{\Phi^{-1}}\rangle$ *for every test function* φ, *where* J_Φ *is the Jacobian determinant of* Φ.

PROOF. Away from the origin $\mathcal{K}\circ\Phi$ is the C^∞ function $K(\Phi(\mathbf{z}))$. Since $\Phi(0) = 0$ it follows that $|\Phi(\mathbf{z})| \gtrsim c|\mathbf{z}|$ for small $|\mathbf{z}|$, and hence $\mathcal{K}\circ\Phi$ satisfies (13.6) on the unit ball because K does. Moreover the assumption $|\Phi(\mathbf{z})| \geq c|\mathbf{z}|^\delta$ for $|\mathbf{z}| \geq 1$, implies the rapid decrease at infinity of $K(\Phi(\mathbf{z}))$, since the same holds for $K(\mathbf{z})$. It remains to verify the cancellation conditions for $\mathcal{K}\circ\Phi$, *i.e.* $|\langle \mathcal{K}\circ\Phi, \varphi(R\cdot)\rangle| \leq C$ for all $R \geq 1$ wherever φ is a normalized C^∞ function supported in the unit ball. However $|\langle \mathcal{K}\circ\Phi, \varphi(R\cdot)\rangle| = \langle \mathcal{K}, \Psi_R\rangle$, where it is easy to verify that for each R, $\Psi_R(\mathbf{x})$ is supported in the ball $|\mathbf{x}| \leq cR^{-1}$, and satisfies $|\partial_{\mathbf{x}}^\alpha \Psi_R(\mathbf{x})| \leq c_\alpha R^{|\alpha|}$, with bounds c and c_α independent of R. Thus we may think of $\Psi_R(\mathbf{x})$ as of the form $c_1\Psi(c_2R\mathbf{x})$, for some c_1 and c_2 independent of R, verifying that $\mathcal{K}\circ\Phi$ satisfies the requisite cancellation condition. Thus the proof of Lemma 13.3 is complete. $\square$

Hence we have proved the representation (b) of Proposition 13.2. The representation (c) is proved in the same way.

13.2. Proof of Theorem 13.1

Turning to the proof of Theorem 13.1, we see by the identity (b) in Proposition 13.2 that the pseudo-differential operator P belongs to the extended class of operators, establishing the first assertion of Theorem 13.1. Now let T be any operator in the extended class,

$$T(f)(\mathbf{x}) = \int \mathcal{L}(\mathbf{x}, \mathbf{y}^{-1}\mathbf{x})f(\mathbf{y})d\mathbf{y},$$

with $\mathcal{L}(\mathbf{x},\mathbf{z})$ a kernel of the extended class. We examine the commutator $[T,P]$. Let B be a ball containing the $\mathbf{x}$–support of the symbol $a(\mathbf{x},\boldsymbol{\xi})$ of P, and hence containing the $\mathbf{x}$-supports of $\mathcal{K}(\mathbf{x},\mathbf{z})$, $\mathcal{K}^L(\mathbf{x},\mathbf{z})$ and $\mathcal{K}^R(\mathbf{x},\mathbf{z})$. We now first decompose T as T_0+T_∞ via $\mathcal{L}(\mathbf{x},\mathbf{z}) = \mathcal{L}_0(\mathbf{x},\mathbf{z})+\mathcal{L}_\infty(\mathbf{x},\mathbf{z})$ where $\mathcal{L}_0(\mathbf{x},\mathbf{z}) = \eta(\mathbf{x})\mathcal{L}(\mathbf{x},\mathbf{z})$, $\mathcal{L}_\infty(\mathbf{x},\mathbf{z}) = (1-\eta(\mathbf{x}))\mathcal{L}_\infty(\mathbf{x},\mathbf{z})$ with $\eta \in C^\infty$ of compact support which $= 1$ on the double B^* at the ball B. Thus $\mathcal{K}_0(\mathbf{x},\mathbf{z})$ has compact $\mathbf{x}$–support, while $\mathcal{K}_\infty(\mathbf{x},\mathbf{z})$ is supported on the complement of B^*. (Note that this is different from the decomposition $T = T_0 + T_\infty$ in Section 12.1).

Now by reasoning as in Section 12.1, we can write

$$T_0(f)(\mathbf{x}) = \sum_n T_n(f)(\mathbf{x}) = \sum_n a_n(\mathbf{x})(f * \mathcal{H}_n)(\mathbf{x}) \tag{13.9}$$

where $a_n(x)$ are C^∞ functions, supported in a common compact set, for which $\sup_{\mathbf{x}} |\partial_{\mathbf{x}}^\alpha a_n(\mathbf{x})|$ is rapidly decreasing in n, for each fixed α. The kernels $\mathcal{H}_n$ are proper kernels and their semi-norms are each uniformly bounded in n.

For the pseudo-differential operator P we shall use the "right–invariant" representation (c) in Proposition 13.2. This gives by the reasoning establishing (13.9), that

$$P(f)(\mathbf{x}) = \sum_{n'} P_{n'}(f)(\mathbf{x}) = \sum_{n'} b_{n'}(\mathbf{x})(\mathcal{K}_{n'} * f)(\mathbf{x}) . \tag{13.10}$$

Here the $b_{n'}(\mathbf{x})$ are C^∞ function, supported in a compact set (in view of the support property of $\mathcal{K}^R(\mathbf{x},\mathbf{z})$), for which $\sup_{\mathbf{x}} |\partial_{\mathbf{x}}^\alpha b_{n'}(\mathbf{x})|$ is rapidly decreasing in n' for each fixed α, and the $\mathcal{K}_{n'}$ are isotropic proper kernels and their semi-norms are each uniformly bounded in n'. The rapid decrease in n and n' pointed out above reduces the study of the commutator $[T_0, P]$ to that where only a single term of each of the sums, (13.9) and (13.10) appears.

Thus we look at the commutator $[a \circ U, b \circ V]$, where $U(f) = f * \mathcal{H}$ and $V(f) = \mathcal{K} * f$. This commutator is handled by the following

LEMMA 13.4.

(1) *U and V commute,*
(2) *$[b, U]$ is a smoothing operator,*
(3) *$[a, V]$ is a smoothing operator,*

and the bounds of the mappings $[b, U]$ and $[a, V]$ depend only on finitely many of the semi-norms controlling $a, b, \mathcal{H}$, and $\mathcal{K}$.

Suppose first that the lemma is proved. Then since $U \circ V = V \circ U$ and $ab = ba$ we have

$$\begin{aligned}
[a \circ U, b \circ V] &= a \circ U \circ b \circ V - b \circ V \circ a \circ U \\
&= ab \circ U \circ V - ba \circ V \circ U + a \circ [U, b] \circ V - b \circ [V, a] \circ U \\
&= a \circ [U, b] \circ V + b \circ [a, V] \circ U
\end{aligned}$$

As well as multiplication by a and b, both U and V map L^p to itself, since they are convolution operators with proper kernels. Thus by conclusions (2) and (3) of the lemma, $[a \circ U, b \circ V]$ maps L^p to L^p_1, and we have achieved our desired result for $[T_0, P]$.

PROOF OF LEMMA 13.4 . The first conclusion (which is key) follows immediately because U is left-invariant and V is right-invariant. Let us now consider $[b, U]$, and begin with the representation given for $[b, U]$ in (12.7) (here $\mathcal{H}$ plays the role of $\mathcal{K}$). We have that $[b, U]$ is an operator of the extended class, represented by the kernel $\widetilde{\mathcal{H}}(\mathbf{x},\mathbf{z}) = c(\mathbf{x},\mathbf{z})\mathcal{H}(\mathbf{x},\mathbf{z})$, where $c(\mathbf{x},\mathbf{z})$ is the C^∞ function given by $b(\mathbf{x}) - b(\mathbf{x}\mathbf{z}^{-1})$. Notice however that $c(\mathbf{x}, 0) = 0$, and thus

$$c(\mathbf{x},\mathbf{z}) = \sum_{k=1}^{n} \sum_{j=1}^{c_k} z_{k,j} \cdot c_{k,j}(\mathbf{x},\mathbf{z}),$$

where $c_{k,j}$ are again C^∞ functions. Hence

$$\widetilde{\mathcal{H}}(\mathbf{x},\mathbf{z}) = \sum_{k,j} z_{k,j} \cdot \widetilde{\mathcal{H}}_{k,j}(\mathbf{x},\mathbf{z}) \tag{13.11}$$

where by Lemma 12.4 we know that each $\widetilde{\mathcal{H}}_{k,j}(\mathbf{x},\mathbf{z})$ is an extended kernel. Now if X is a left-invariant vector field, then

$$X[b,U](f)(\mathbf{x}) = X\left(\int \widetilde{\mathcal{H}}(\mathbf{x},\mathbf{y}^{-1}\mathbf{x})f(\mathbf{y})d\mathbf{y}\right) = \int \mathcal{H}'(\mathbf{x},\mathbf{y}^{-1}\mathbf{x})f(\mathbf{y})d\mathbf{y} ,$$

where by (13.11) we have that

$$\mathcal{H}'(\mathbf{x},\mathbf{z}) = \sum_{j,k=1}^{n} \Big(X_{\mathbf{x}}\big(z_{k.j} \cdot \widetilde{\mathcal{H}}_{k,j}(\mathbf{x},\mathbf{z})\big) + X_{\mathbf{z}}\big(z_{k,j} \cdot \widetilde{\mathcal{H}}_{k,j}(\mathbf{x},\mathbf{z})\big)\Big) .$$

At this stage we need the following observation.

LEMMA 13.5. *Suppose $\mathcal{H}$ is a proper kernel and X is a left-invariant vector field of degree 1 (that is, in the subspace $\mathfrak{h}_1$ of the Lie algebra). Then for each (k,j), $1 \le k \le n$, $1 \le j \le c_k$, $X\big(z_{k,j}\mathcal{H}(\mathbf{z})\big)$ is a proper kernel.*

This can be proved by using the Fourier transform characterization of proper kernels. First write X as

$$X = \frac{\partial}{\partial z_{1,r}} + \sum_{\ell=2}^{n}\sum_{s=1}^{c_\ell} q_{\ell,s}(\mathbf{z}) \frac{\partial}{\partial z_{\ell,s}} = \frac{\partial}{\partial z_{1,r}} + \sum_{\ell=2}^{n} Q_\ell(\mathbf{z}) \cdot \nabla_{\mathbf{z}_\ell}$$

where $Q_\ell(\mathbf{z}) = Q_\ell(\mathbf{z}_1, \cdots, \mathbf{z}_{\ell-1})$ are homogeneous polynomials of degree $\ell - 1$ (in the homogeneity of the automorphic dilations). Consider first the case $k = 1$. The Fourier transform of $X\big(z_{1,j}\mathcal{H}(\mathbf{z})\big)$ is then a linear combination of $m = \widehat{\mathcal{H}}$ and $\boldsymbol{\xi}_\ell \cdot Q_\ell(2\pi i \partial_{\boldsymbol{\xi}})\, \partial_{\xi_{1,j}} m(\boldsymbol{\xi})$. However, it is clear from (13.1) that $\widehat{N}_j(\boldsymbol{\xi})^k \ge |\boldsymbol{\xi}_k|$ if $j \le k$. The the effect of the factor $\boldsymbol{\xi}_\ell$, which has size $|\boldsymbol{\xi}_\ell|$, is counterbalanced by the effect of $Q_\ell(2\pi i \partial_{\boldsymbol{\xi}})\partial_{\xi_{k,j}}$ on $m(\boldsymbol{\xi})$, in view of the estimates for m. The situation when $k > 1$ (and hence $\partial_{\xi_{1,j}}$ is replaced by $\partial_{\xi_{k,j}}$) leads to an even better estimate, which then establishes Lemma 13.5.

Using the lemma we see therefore that $X_{\mathbf{x}}[b,U]$ is an operator with an extended kernel, and hence by Theorem 12.1 we have that $[b,U]$ maps L^p to L^p_1, establishing conclusion (2) of Lemma 13.4.

Finally, for conclusion (3) we may appeal to the standard theory of pseudo-differential operators that guarantees that the commutator $[a,V]$ is an operator of "order 1", and in particular maps L^p to L^p_1; or we may argue in the same spirit as in the case of conclusion (2), but here the details are simpler. In any case, Lemma 7.4 is now established and this proves that $[T_0, P]$ is bounded from L^p to L^p_1. □

To complete the proof of Theorem 13.1 it remains to deal with $[T_\infty, P] = T_\infty P - PT_\infty$. We will see that the terms $T_\infty P$ and PT_∞ are separately smoothing operators (in fact, infinitely smoothing) because of disjointness of relevant supports. Consider first $T_\infty P$. Then $T_\infty(Pf) = T_\infty(F)$, where F is supported in the ball B (where the symbol $a(\mathbf{x},\boldsymbol{\xi})$ is supported). However

$$T_\infty(F)(\mathbf{x}) = \int \mathcal{K}_\infty(\mathbf{x}, \mathbf{y}^{-1}\mathbf{x})F(\mathbf{y})d\mathbf{y}$$

and $\mathcal{K}_\infty(\mathbf{x},\mathbf{z})$ is supported for $\mathbf{x}$ outside B^*. Hence in the integral above we have that $|\mathbf{y}^{-1}\mathbf{x}| \ge c > 0$, for an appropriate c. Moreover if X is any left–invariant operator then $X(T_\infty F)(\mathbf{x}))$ is given by $\int \mathcal{K}'_\infty(\mathbf{x},\mathbf{y}^{-1}\mathbf{x})F(\mathbf{y})d\mathbf{y}$ with $K'_\infty(\mathbf{x},\mathbf{z}) = X_{\mathbf{x}}\mathcal{K}_\infty(\mathbf{x},\mathbf{z}) + X_{\mathbf{z}}\mathcal{K}_\infty(\mathbf{x},\mathbf{z})$. Since now $\mathcal{K}'_\infty(\mathbf{x},\mathbf{z})$ is bounded and has rapid decay in $\mathbf{z}$, (we can restrict ourselves here to $|\mathbf{z}| \ge c > 0$), this ensures that $T_\infty P$ maps L^p to L^p_1. In fact the same argument shows that $X^\alpha(T_\infty P)$ maps L^p to itself for any monomial X^α of left–invariant differential operators.

Essentially the same argument works for PT_∞. Here

$$(PT_\infty)(f) = \int K^L(\mathbf{x}, \mathbf{y}^{-1}\mathbf{x})f(\mathbf{y})d\mathbf{y}$$

by Proposition 13.2, with F supported in the complement of B^* and $\mathcal{K}(\mathbf{x},\mathbf{z})$ supported where $\mathbf{x} \in B$. Then again $|\mathbf{y}^{-1}\mathbf{x}| \geq c > 0$, and we may proceed as before. Altogether then, the proof of Theorem 13.1 is completed.

13.3. The space L_1^p

We conclude by proving the equivalence of the condition that $f \in L_1^p$ with the Lipschitz–type condition (13.5).

PROPOSITION 13.6. Let $1 < p < \infty$. Then

$$f \in L_1^p \qquad \Longleftrightarrow \qquad ||f(\mathbf{xh}) - f(\mathbf{x})||_{L_p} = O\big(N_n(\mathbf{h})\big), \text{ as } h \to 0.$$

PROOF. Let X be a left-invariant vector field in the subspace $\mathfrak{h}_1$ of the Lie algebra, and set $\mathbf{h}_t = \exp(tX)$. Then $|\mathbf{h}_t| = |\mathbf{h}_t| \approx |t|,\ t \in \mathbb{R}$. So

$$\begin{aligned} &||f(\mathbf{xh}) - f(\mathbf{x})||_{L_p} = O\big(N_n(\mathbf{h})\big), \quad \text{as } h \to 0 \\ &\qquad \Longrightarrow \qquad ||\frac{1}{t}\big(f(\mathbf{xh}_t) - f(\mathbf{x})\big)||_{L^p} = O(1) \quad \text{as } t \to 0. \end{aligned}$$

Thus by the weak compactness of L^p, there is a sequence $\{t_n\},\ t_n \to 0$ so that $\frac{1}{t_n}\big(f(\mathbf{xh}_{t_n}) - f(\mathbf{x})\big)$ converges weakly in L^p and hence in the sense of distributions. Thus in that sense, $X(f) \in L^p$. Since this holds for all $X \in \mathfrak{h}_1$, we have that $f \in L_1^p$.

Conversely, suppose $f \in L_1^p$, and $\mathbf{h} = \exp(sX)$ where $X \in \mathfrak{h}_1$. Then $N_n(\mathbf{h}) = |\mathbf{h}| \lesssim |s|$ and

$$f(\mathbf{xh}) - f(\mathbf{x}) = \int_0^s \frac{d}{dt} f\big(\mathbf{x}\exp(tX)\big)\, dt = \int_0^s Xf\big(\mathbf{x}\exp(tX)\big)\, dt\ .$$

Hence, $||f(\mathbf{xh}) - f(\mathbf{x})||_{L^p} \leq |s|\ ||X(f)||_{L^p} = O\big(N_n(\mathbf{h})\big)$. To extend this to any element $\mathbf{h}$ we use the following assertion.

LEMMA 13.7. *Suppose G is a stratified group. Then there is an integer M, so that every $\mathbf{h} \in G$, can be written as $\mathbf{h} = \mathbf{h}_1\mathbf{h}_2 \cdots \mathbf{h}_M$, where each $\mathbf{h}_j \in \exp\mathfrak{h}_1$ and $|\mathbf{h}_j| \leq cN_n(\mathbf{h})$ for all $1 \leq j \leq M$.*

Assuming for a moment this to be true, the desired estimate for $f(\mathbf{xh}) - f(\mathbf{x})$ then follows for the special case when $\mathbf{h} \in \exp\mathfrak{h}_1$ by writing

$$f(\mathbf{xh}) - f(\mathbf{x}) = \sum_{j=1}^{M} f(\mathbf{xh}_1 \cdots \mathbf{h}_j) - f(\mathbf{xh}_1 \cdots \mathbf{h}_{j-1}). \qquad \square$$

PROOF OF LEMMA 13.7. Let $V = \exp(B_1)$, where B_1 is the unit ball in $\mathfrak{h}_1$, Then V is an analytic submanifold of G which generates G. By Proposition 1.1 of [**RS88**], there exists an integer m such that V^m contains an open set A. Since $V = V^{-1}$, V^{2m} contains a neighborhood of the identity. This proves the statement for $|\mathbf{h}| < \delta$ for some $\delta > 0$ with $M = 2m$. By homogeneity, the same holds for every $\mathbf{h} \in G$. $\square$

CHAPTER 14

Appendix I: Properties of cones $\Gamma(\mathbf{A})$

In this appendix we study properties of the cone $\Gamma(\mathbf{E})$ defined in (5.1). Our results provide motivation for the basic hypothesis (2.10) imposed on the matrix $\mathbf{E}$.

14.1. Optimal inequalities and the basic hypothesis

DEFINITION 14.1. Let $\mathbf{A} = \{a(j,k) : 1 \leq j,k \leq n\}$ be an $n \times n$ matrix of strictly positive real numbers. Then

$$\Gamma^o(\mathbf{A}) = \Big\{\mathbf{t} = (t_1, \dots, t_n) \in \mathbb{R}^n : a(j,k)t_k < t_j < 0\Big\}$$
$$\Gamma(\mathbf{A}) = \Big\{\mathbf{t} = (t_1, \dots, t_n) \in \mathbb{R}^n : a(j,k)t_k \leq t_j < 0\Big\}$$

are (possibly empty) convex polyhedral cones contained in the negative orthant of $\mathbb{R}^n$. The dimension of either cone is the maximal number of linearly independent vectors it contains. Equivalently, this is the dimension of the smallest subspace of $\mathbb{R}^n$ containing the cone.

$\Gamma(\mathbf{A})$ is defined by a collection of linear inequalities involving pairs of coordinates. It may happen that different matrices (*i.e.* different sets of such inequalities) give rise to the same cone. The following Lemma shows that there is a unique optimal set of inequalities, and that the corresponding matrix satisfies the basic hypothesis (2.10).

LEMMA 14.2. *Let $\mathbf{A} = \{a(j,k)\}$ be an $n \times n$ matrix of strictly positive real numbers, and suppose that the corresponding cone $\Gamma(\mathbf{A})$ is not empty. Then there exists a unique $n \times n$ matrix $\widetilde{\mathbf{A}} = \{\tilde{a}(j,k)\}$ with the following properties:*

(a) *The coefficients of $\widetilde{\mathbf{A}}$ satisfy the inequalities (2.10) of the basic hypothesis:*
$$\begin{aligned} 1 &= \tilde{a}(j,j) && \text{for } 1 \leq j \leq n, \\ 0 &< \tilde{a}(j,k) \leq \tilde{a}(j,l)\tilde{a}(l,k) && \text{for } 1 \leq j,k,l \leq n. \end{aligned}$$

(b) $\Gamma(\mathbf{A}) = \Gamma(\widetilde{\mathbf{A}})$.

Moreover, if $\widetilde{\mathbf{A}}$ satisfies (a) *and* (b), *then $\tilde{a}(j,k) \leq a(j,k)$ for $1 \leq j,k \leq n$.*

PROOF. For any $\mathbf{t} = (t_1, \dots, t_n) \in \Gamma(\mathbf{A})$, $t_j < 0$ and $0 < t_j t_k^{-1} \leq a(j,k)$ for $1 \leq j,k \leq n$. If $\Gamma(\mathbf{A}) \neq \emptyset$ define

$$0 < \tilde{a}(j,k) = \sup_{\mathbf{t} \in \Gamma(\mathbf{A})} \frac{t_j}{t_k} \leq a(j,k). \tag{14.1}$$

Then $\tilde{a}(j,j) = 1$ and for any $j,k,l \in \{1, \dots, n\}$

$$\tilde{a}(j,k) = \sup_{\mathbf{t} \in \Gamma(\mathbf{A})} \frac{t_j}{t_k} = \sup_{\mathbf{t} \in \Gamma(\mathbf{A})} \left[\frac{t_j}{t_l}\frac{t_l}{t_k}\right] \leq \sup_{\mathbf{t} \in \Gamma(\mathbf{A})} \frac{t_j}{t_l} \sup_{\mathbf{t} \in \Gamma(\mathbf{A})} \frac{t_l}{t_k} = \tilde{a}(j,l)\,\tilde{a}(l,k).$$

This establishes (a).

Let $\mathbf{v} = (v_1, \ldots, v_n) \in \Gamma(\widetilde{\mathbf{A}})$. Since $v_k < 0$ and $\tilde{a}(j,k) \leq a(j,k)$, it follows that $a(j,k)v_k \leq \tilde{a}(j,k)v_k \leq v_j$, and so $v_j v_k^{-1} \leq a(j,k)$. Thus $\mathbf{v} \in \Gamma(\mathbf{A})$ and it follows that $\Gamma(\widetilde{\mathbf{A}}) \subset \Gamma(\mathbf{A})$. Conversely, if $\mathbf{v} = (v_1, \ldots, v_n) \in \Gamma(\mathbf{A})$, then $v_j v_k^{-1} \leq \sup_{\mathbf{t} \in \Gamma(\mathbf{A})} t_j t_k^{-1} = \tilde{a}(j,k)$ and so $\tilde{a}(j,k)v_k \leq v_j$. It follows that $\Gamma(\mathbf{A}) \subset \Gamma(\widetilde{\mathbf{A}})$, and this establishes (b).

Next we show that the matrix $\widetilde{\mathbf{A}}$ is unique. Let $\mathbf{C} = \{c(j,k) : 1 \leq j,k \leq n\}$ be any matrix satisfying (a) and (b). Since $\Gamma(\mathbf{C}) = \Gamma(\mathbf{A})$, if $\mathbf{t} \in \Gamma(\mathbf{C})$ then $c(j,k)t_k \leq t_j$, and so

$$\tilde{a}(j,k) = \sup_{\mathbf{t} \in \Gamma(\mathbf{A})} t_j t_k^{-1} = \sup_{\mathbf{t} \in \Gamma(\mathbf{C})} t_j t_k^{-1} \leq c(j,k).$$

To establish the reverse inequality, consider the vectors

$$\mathbf{w}_l = \big(-c(1,l), \ldots, -c(n,l)\big).$$

Since $\mathbf{C}$ satisfies (a), we have $c(j,l) \leq c(j,k)c(k,l)$, or equivalently $c(j,k)\big(-c(k,l)\big) \leq \big(-c(j,l)\big)$. This shows that $\mathbf{w}_l \in \Gamma(\mathbf{C}) = \Gamma(\widetilde{\mathbf{A}})$ for $1 \leq l \leq n$. It follows that $\tilde{a}(j,k)(-c(k,l)) \leq -c(j,l)$ for all $1 \leq j,k,l \leq n$. Letting $l = k$, we have $\tilde{a}(j,k)(-c(k,k)) \leq -c(j,k)$, and since $c(k,k) = 1$ we have $c(j,k) \leq \tilde{a}(j,k)$. Thus $c(j,k) = \tilde{a}(j,k)$, which shows that $\widetilde{\mathbf{A}}$ is unique. We have already observed that $\tilde{a}(j,k) \leq a(j,k)$, and since the matrix $\widetilde{\mathbf{A}}$ is unique, this completes the proof. □

It can be shown that, in the hypotheses of Lemma 14.2,

$$\tilde{a}(j,k) = \begin{cases} \min\{a(j,i_1)a(i_1,i_2)\cdots a(i_r,k) : r \geq 0, \{i_1,\ldots,i_r\} \subset \{1,\ldots,n\}\} & \text{if } j \neq k \\ 1 & \text{if } j = k \end{cases}.$$

14.2. Partial matrices

In Lemma 14.2 it is assumed that the cone $\Gamma(\mathbf{A})$ is defined by the full number $n(n-1)$ inequalities $a(j,k)t_k \leq t_j$ where $1 \leq j \neq k \leq n$. We also need to consider cases in which we begin with fewer inequalities. We formulate this as follows. Let $\mathbf{B} = \{b(j,k)\}$ be an $n \times n$ matrix, where this time each $b(j,k)$ is either a positive real number or the symbol ∞. As before we put

$$\Gamma(\mathbf{B}) = \Big\{\mathbf{t} = (t_1, \ldots, t_n) \in \mathbb{R}^n : b(j,k)t_k \leq t_j < 0\Big\},$$

but now if $b(j,k) = \infty$, the inequality $b(j,k)t_k \leq t_j < 0$ puts *no* constraint on the relation between the two negative real numbers t_k and t_j. It is still the case that $\Gamma(\mathbf{B})$ is a polyhedral cone in the negative orthant of $\mathbb{R}^n$. We call $\mathbf{B}$ a *partial matrix*. We say that a partial matrix $\mathbf{B} = \{b(j,k)\}$ is *connected* if for any two indices $j,k \in \{1,\ldots,n\}$ there are indices $i_1,\ldots,i_a \in \{1,\ldots,n\}$ (with a possibly dependent on j and k) so that

$$b(j,i_1)b(i_1,i_2)\cdots b(i_l,i_{l+1})\cdots b(i_{a-1},i_a)b(i_a,k) < +\infty.$$

It then follows that

$$b(j,i_1)b(i_1,i_2)\cdots b(i_j,i_{j+1})\cdots b(i_{a-1},i_a)b(i_a,k)t_k \leq t_j < 0$$

so if $\mathbf{B}$ is connected and $\mathbf{t} \in \Gamma(\mathbf{B})$ it follows that

$$\sup_{\mathbf{t}\in\Gamma(\mathbf{B})} \frac{t_j}{t_k} \leq b(j, i_1)b(i_1, i_2)\cdots b(i_j, i_{j+1})\cdots b(i_{a-1}, i_a)b(i_a, k) < +\infty.$$

We then have the following analogue of Lemma 14.2.

LEMMA 14.3. *Let $\mathbf{B} = \{b(j,k)\}$ be an $n \times n$ connected, partial matrix, and suppose that the corresponding cone $\Gamma(\mathbf{B})$ is not empty. Then there exists a unique $n \times n$ matrix $\widetilde{\mathbf{B}} = \{\tilde{b}(j,k)\}$ with the following properties:*

(a) *The coefficients of $\widetilde{\mathbf{B}}$ are all finite positive real numbers, and*

$$\begin{aligned} 1 &= \tilde{b}(j,j) && \text{for } 1 \leq j \leq n, \\ 0 < \tilde{b}(j,k) &\leq \tilde{b}(j,l)\tilde{b}(l,k) && \text{for } 1 \leq j,k,l \leq n. \end{aligned}$$

(b) $\Gamma(\mathbf{B}) = \Gamma(\widetilde{\mathbf{B}})$.

Moreover, if $\widetilde{\mathbf{B}}$ satisfies (a) *and* (b)*, then $\tilde{b}(j,k) \leq b(j,k)$ for $1 \leq j,k \leq n$.*

PROOF. There is only a minor change from the proof of Lemma 14.2. Since $b(j,k)$ might equal ∞ we cannot directly conclude that $\sup_{\mathbf{t}\in\Gamma(\mathbf{B})} t_j\, t_k^{-1}$ is finite by observing that it is bounded by $b(j,k)$ as in equation (14.1). However, since $\mathbf{B}$ is connected, it is still the case that $\sup_{\mathbf{t}\in\Gamma(\mathbf{B})} t_j\, t_k^{-1} < \infty$, and we can define $\widetilde{b}(j,k)$ to the this supremum. The proof of (a) then proceeds as before, and we set $\Gamma(\widetilde{\mathbf{B}}) = \Big\{\mathbf{t} = (t_1,\ldots,t_n) \in \mathbb{R}^n : \widetilde{b}(j,k)t_k \leq t_j < 0\Big\}$. Note that if $b(j,k) < \infty$ then $\widetilde{b}(j,k) = \sup_{\mathbf{t}\in\Gamma(\mathbf{B})} \frac{t_j}{t_k} \leq b(j,k)$ as before. The proofs that $\Gamma(\widetilde{\mathbf{B}}) = \Gamma(\mathbf{B})$ and the uniqueness of $\mathbf{B}$ also follow as before. □

14.3. Projections

Let $1 < p < n$ and let $\pi_{n,p} : \mathbb{R}^n \to \mathbb{R}^p$ be the projection onto the first p coordinates: $\pi_{n,p}(x_1,\ldots,x_n) = (x_1,\ldots,x_p)$. Let $\mathbf{A}_n = \{a(j,k) : 1 \leq j,k \leq n\}$ be an $n \times n$ matrix with strictly positive entries which satisfies the basic hypotheses

$$\begin{aligned} a(j,j) &= 1 && 1 \leq j \leq n, \\ a(j,l) &\leq a(j,k)a(k,l) && 1 \leq j,k,l \leq n. \end{aligned}$$

Let $\mathbf{A}_p = \{a(j,k) : 1 \leq j,k \leq p\}$ be the corresponding $p \times p$ matrix (which of course also satisfies the basic hypotheses). We than have two cones:

$$\Gamma(\mathbf{A}_n) = \Big\{\mathbf{t} = (t_1,\ldots,t_n) \in \mathbb{R}^n : a(j,k)t_k \leq t_j < 0,\ 1 \leq j,k \leq n\Big\},$$

$$\Gamma(\mathbf{A}_p) = \Big\{\mathbf{u} = (u_1,\ldots,u_p) \in \mathbb{R}^p : a(j,k)u_k \leq u_j < 0,\ \leq j,k \leq p\Big\}.$$

LEMMA 14.4. *The projection $\pi_{n,p}$ maps the cone $\Gamma(\mathbf{A}_n)$ onto the cone $\Gamma(\mathbf{A}_p)$.*

PROOF. If $a(j,k)t_k \leq t_j < 0$ for $1 \leq j,k \leq n$, then $a(j,k)t_k \leq t_j < 0$ for $1 \leq j,k \leq p$, and so it is clear that π maps $\Gamma(\mathbf{A}_n)$ into $\Gamma(\mathbf{A}_p)$. The main point of the Lemma is that the mapping is *onto*. But since

$$\pi_{n,p} = \pi_{p+1,p} \circ \pi_{p+2,p+1} \circ \cdots \circ \pi_{n,n-1}$$

it clearly suffices to show that the mapping $\pi_{n,n-1} : \Gamma(\mathbf{A}_n) \to \Gamma(\mathbf{A}_{n-1})$ is onto.

Let $(u_1,\ldots,u_{n-1}) \in \Gamma(\mathbf{A}_{n-1})$. We must show that there exists $u_n < 0$ so that

$$1 \leq j \leq n-1 \quad \Longrightarrow \quad a(j,n)u_n \leq u_j \quad \text{and} \quad a(n,j)u_j \leq u_n$$

If we let E_j be the closed interval $\left[a(n,j)u_j, \frac{u_j}{a(j,n)}\right]$, then we must show that $\bigcap_{j=1}^{n-1} E_j \neq \emptyset$. For any $j,k \in \{1,\ldots,n-1\}$ we have

$$\begin{aligned} a(n,j)u_j &\leq \frac{a(k,j)}{a(k,n)}u_j && \text{since } a(k,j) \leq a(k,n)a(n,j) \text{ and } u_j < 0, \\ &\leq \frac{u_k}{a(k,n)} && \text{since } (u_1,\ldots,u_{n-1}) \in \Gamma(\mathbf{A}_{n-1}) \text{ so } a(k,j)u_j \leq u_k. \end{aligned}$$

It follows that for any index $1 \leq \ell \leq n-1$ we have

$$\begin{aligned} a(n,\ell)u_\ell &\leq \sup\Big\{a(n,j)u_j : 1 \leq j \leq n-1\Big\} \\ &\leq \inf\Big\{\frac{u_k}{a(k,n)} : 1 \leq k \leq n-1\Big\} \\ &\leq \frac{u_\ell}{a(\ell,n)} \end{aligned}$$

and hence we have produce a closed, non-empty interval in $\bigcap_{j=1}^{n-1} E_j$:

$$\left[\sup\Big\{a(n,j)u_j : 1 \leq j \leq n-1\Big\}, \inf\Big\{\frac{u_k}{a(k,n)} : 1 \leq k \leq n-1\Big\}\right] \subset \bigcap_{j=1}^{n-1} E_j.$$

□

14.4. The dimension of $\Gamma(\mathbf{E})$

PROPOSITION 14.5. Let $\mathbf{E}$ be an $n \times n$ matrix satisfying the basic hypothesis (2.10). Then the cone $\Gamma(\mathbf{E})$ is not empty. In particular, the $2n$ vectors

$$\begin{aligned} \mathbf{v}_l &= \big(-e(1,l),\ldots,-e(n,l)\big), && 1 \leq l \leq n \\ \mathbf{w}_l &= \big(-e(l,1)^{-1},\ldots,-e(l,n)^{-1}\big), && 1 \leq l \leq n \end{aligned}$$

all belong to $\Gamma(\mathbf{E})$.

PROOF. Fix $l \in \{1,\ldots,n\}$ and let $v_j = -e(j,l)$. Then it follows from (2.10) that $e(j,l) \leq e(j,k)e(k,l)$ and so $e(j,k)v_k = -e(j,k)e(k,l) \leq -e(j,l) = v_j$. This shows that $\mathbf{v}_l \in \Gamma(\mathbf{E})$. Similarly, let $w_j = -e(l,j)^{-1}$. Then since $e(l,k) \leq e(l,j)e(j,k)$, $e(j,k)w_k = -\frac{e(j,k)}{e(l,k)} \leq -\frac{1}{e(l,j)}$, and so $\mathbf{w}_l \in \Gamma(\mathbf{E})$. This completes the proof. □

Since the vectors $\mathbf{v}_l$ in Proposition 14.5 are the negatives of the columns of $\mathbf{E}$, we have the following corollary.

COROLLARY 14.6. Let $\mathbf{E}$ be an $n \times n$ matrix satisfying the basic hypothesis (2.10). Then the dimension of $\Gamma(\mathbf{E})$ is greater than or equal to the rank of $\mathbf{E}$.

LEMMA 14.7. *Let $\mathbf{E}$ be an $n \times n$ matrix satisfying the basic hypotheses* (2.10). *The following are equivalent:*

(i) *the interior of $\Gamma(\mathbf{E})$ is non-empty;*
(ii) *the dimension of $\Gamma(\mathbf{E})$ is n;*
(iii) $1 < e(j,k)e(k,j)$ *for every pair $1 \leq j \neq k \leq n$.*

PROOF. Obviously (i) implies (ii). Conversely, since $\Gamma(\mathbf{E})$ is convex, if it contains a basis of $\mathbb{R}^n$ it also containes the convex hull of this basis together with 0, which is open.

In order to prove that (iii) implies (i), we argue by induction on n. For $n = 1$, $\Gamma(\mathbf{E}) = (-\infty, 0)$ which has dimension 1. For $n = 2$, $\det(\mathbf{E}) = 1 - e(1,2)e(2,1) < 0$, so the rank of $\mathbf{E}$ is 2, and hence by Corollary 14.6 the dimension of $\Gamma(\mathbf{E})$ is 2 and its interior is a nontrivial angle.

Thus assume that the Lemma is true for all $(n-1) \times (n-1)$ matrices $\mathbf{E}'$ satisfying equation (2.10). Let $\mathbf{E}$ be an $n \times n$ matrix satisfying equation (2.10) such that $1 < e(j,k)e(k,j)$ for every pair $1 \leq j \neq k \leq n$. Let $\mathbf{E}'$ be the submatrix consisting of the first $n-1$ rows and columns. Then clearly $\mathbf{E}'$ also satisfies equation (2.10) and the hypothesis of the Lemma, so by the induction hypothesis, the dimension of $\Gamma(\mathbf{E}')$ is $n-1$, and $\Gamma(\mathbf{E}')$ contains a non-empty open set $U \subset \mathbb{R}^{n-1}$.

Let $\mathbf{t}' = (t_1, \ldots, t_{n-1}) \in U$. Then the components of $\mathbf{t}'$ satisfy the strict inequalities $e(k,j)t_j < t_k$ for all $j \neq k$. Since $t_j < 0$ and $e(k,j) \leq e(k,n)e(n,j)$, it follows that $e(k,n)e(n,j)t_j \leq e(k,j)t_j < t_k$ and so $e(n,j)t_j < e(k,n)^{-1}t_k$ for all j, k. Therefore

$$\max\Big\{e(n,j)t_j : 1 \leq j \leq n-1\Big\} < \min\Big\{e(k,n)^{-1}t_k : 1 \leq k \leq n-1\Big\}. \tag{14.2}$$

But if $\mathbf{t}' \in \Gamma(\mathbf{E}')$ then $\mathbf{t} = (\mathbf{t}', t_n) = (t_1, \ldots, t_{n-1}, t_n)$ in $\Gamma(\mathbf{E})$ if and only if for all $1 \leq j, k \leq n-1$,

$$\max\Big\{e(n,j)t_j : 1 \leq j \leq n-1\Big\} \leq t_n \leq \min\Big\{e(k,n)^{-1}t_k : 1 \leq k \leq n-1\Big\}.$$

It follows from (14.2) that the set of t_n for which this is true is non empty and contains the open interval $\big(m(\mathbf{t}'), M(t')\big)$ where

$$m(\mathbf{t}') = \max\Big\{e(n,j)t_j : 1 \leq j \leq n-1\Big\},$$
$$M(\mathbf{t}') = \min\Big\{e(k,n)^{-1}t_k : 1 \leq k \leq n-1\Big\}.$$

Then

$$\Big\{(\mathbf{t}', t_n) : \mathbf{t}' \in U \text{ and } m(\mathbf{t}') < t_n < M(\mathbf{t}')\Big\}$$

is an open subset of $\Gamma(\mathbf{E})$.

Conversely, if $1 < e(j,k)e(k,j)$ for every pair $1 \leq j \neq k \leq n$, it follows from Lemma 14.7 in Appendix A that $\Gamma(\mathbf{E})$ has dimension n and in particular has non-empty interior. On the other hand, if $\Gamma^o(\mathbf{E})$ is non-empty, then the projection of $\Gamma(\mathbf{E})$ onto the two dimensional subspace spanned by x_j and x_k is open and non-empty. Since this projection is contained in $\{(x_j, x_k) \in \mathbb{R}^2 : e(j,k)x_k < x_j < e(k,j)^{-1}x_k\}$, it follows that $1 < e(j,k)e(k,j)$. □

PROPOSITION 14.8. Let $\mathbf{E}$ be an $n \times n$ matrix satisfying the basic hypothesis (2.10), and suppose that $1 < e(j,k)e(k,j)$ for all $j \neq k$. Then for any indices $k_1, \ldots, k_s$ it follows that

$$1 < e(k_1,k_2)e(k_2,k_3)\cdots e(k_{s-1},k_s)e(k_s,k_1)$$

provided that the indices $\{k_1, \ldots, k_s\}$ are not all the same.

PROOF. The basic hypothesis implies that we always have

$$1 \leq e(k_1,k_2)e(k_2,k_3)\cdots e(k_{s-1},k_s)e(k_s,k_1).$$

We argue by induction on s. If $s = 2$, the hypothesis of the Proposition implies that we cannot have $1 = e(k_1, k_2)e(k_2, k_1)$ unless $k_1 = k_2$. Now suppose the statement is true for any set of indices of length $s-1$, and let $k_1, \dots, k_s$ be indices which are not all equal. Consider the product $e(k_1, k_2)e(k_2, k_3)\cdots e(k_{s-1}, k_s)e(k_s, k_1)$. Without loss of generality we can assume $k_1 \neq k_2$ for otherwise $e(k_1, k_2) = 1$ and the result follows by induction. But then

$$\begin{aligned} 1 < e(k_1, k_2)e(k_2, k_1) &\le e(k_1, k_2)e(k_2, k_3)e(k_3, k_1) \le \cdots \\ \cdots &\le e(k_1, k_2)e(k_2, k_3)\cdots e(k_{s-1}, k_s)e(k_s, k_1) \end{aligned}$$

which completes the proof. □

14.5. The reduced matrix $\mathbf{E}^\flat$

In this section we show that if the dimension of $\Gamma(\mathbf{E})$ is $s < n$, there is an $s \times s$ sub-matrix $\mathbf{E}^\flat$ of $\mathbf{E}$ satisfying the hypothesis of Lemma 14.7, so that $\Gamma(\mathbf{E}^\flat) \subset \mathbb{R}^s$ has dimension s, and that $\Gamma(\mathbf{E})$ is the graph of a linear mapping from $\mathbb{R}^s$ to $\mathbb{R}^n$.

Let $\mathbf{E} = \{e(j,k)\}$ be an $n \times n$ matrix satisfying the basic hypothesis (2.10). If $j, k \in \{1, \dots, n\}$ write $j \sim k$ if and only if $e(j,k)e(k,j) = 1$. It is immediate that $j \sim j$ and $j \sim k \iff k \sim j$. Suppose that $j \sim k$ and $k \sim l$. Then since $e(j,k)e(k,j) = 1$ and $e(k,l)e(l,k) = 1$,

$$\begin{aligned} 1 \le e(j,l)e(l,j) &\le \big(e(j,k)e(k,l)\big)\,\big(e(l,k)e(k,j)\big) \\ &= \big(e(j,k)e(k,j)\big)\,\big(e(k,l)e(l,k)\big) = 1 \end{aligned}$$

so that $j \sim l$. Thus '$\sim$' is an equivalence relation.

There are other equivalent formulations of this relation, based on the following remarks.

(i) If $e(j,k)e(k,j) = 1$, the basic hypotheses imply that, for every ℓ,

$$\frac{1}{e(k,j)} \le \frac{e(j,\ell)}{e(k,\ell)} \le e(j,k), \qquad \frac{1}{e(j,k)} \le \frac{e(\ell,j)}{e(\ell,k)} \le e(k,j).$$

Hence the above inequalities are all equalities, i.e., the j-th and k-th row of $\mathbf{E}$ are proportional. Conversely, if the j-th and the k-th rows $\mathbf{E}_j, \mathbf{E}_k$ are proportional, then $\mathbf{E}_j = e(j,k)\mathbf{E}_k$ and $\mathbf{E}_k = e(k,j)\mathbf{E}_j$, hence $e(j,k)e(k,j) = 1$.

In the same way, one proves that $j \sim k$ if and only if the j-th and k-th column are proportional.

(ii) If $\mathbf{t} = (t_1, \dots, t_n) \in \Gamma(\mathbf{E})$ and if $j \sim k$, then

$$e(j,k)t_k \le t_j = e(j,k)e(k,j)t_j \le e(j,k)t_k$$

so for all $\mathbf{t} \in \Gamma(\mathbf{E})$,

$$\mathbf{t} = (t_1, \dots, t_n) \in \Gamma(\mathbf{E}),\ j \sim k \qquad \Longrightarrow \qquad t_j = e(j,k)t_k. \tag{14.3}$$

Conversely, if $\Gamma(\mathbf{E})$ is contained in the hyperplane $t_j = \lambda t_k$, this condition is satisfied in particular by the vectors $\mathbf{v}_k, \mathbf{w}_k$ in (14.5). Hence $e(j,k) = \lambda = e(k,j)^{-1}$ and therefore $j \sim k$.

Let $\{k_1, \dots, k_s\}$ be representatives of the distinct equivalence classes in the set $\{1, \dots, n\}$ and let $\mathbf{E}^\flat$ be the $s \times s$ matrix whose entries are $\{e(k_a, k_b) : 1 \le a, b \le s\}$. Then $\mathbf{E}^\flat$ satisfies the same basic hypothesis, and if $a \neq b$ we have $e(k_a, k_b)e(k_b, k_a) > 1$. The matrix $\mathbf{E}^\flat$ is called a *reduced matrix* for $\mathbf{E}$. It is not uniquely determined by $\mathbf{E}$ since it depends on the choice of representatives of the s distinct equivalence classes. The number s is called the *reduced rank* of $\mathbf{E}$.

Let $\Pi : \mathbb{R}^n \to \mathbb{R}^s$ be the projection given by

$$\Pi(\mathbf{t}) = \Pi(t_1, \ldots, t_n) = (t_{k_1}, \ldots, t_{k_s}).$$

If $\mathbf{t} \in \Gamma(\mathbf{E})$ then clearly $\Pi(\mathbf{t}) \in \Gamma(\mathbf{E}^\flat)$, so $\Pi : \Gamma(\mathbf{E}) \to \Gamma(\mathbf{E}^\flat)$. On the other hand, if $\mathbf{t}^\flat = (t_{k_1}, \ldots, t_{k_s}) \in \Gamma(\mathbf{E}^\flat)$, set $\sigma(\mathbf{t}^\flat) = (t_1, \ldots, t_n)$ where $t_j = e(j, k_a)t_{k_a}$ if $j \sim k_a$. We claim $\sigma(\mathbf{t}^\flat) \in \Gamma(\mathbf{E})$. To see this, let $1 \le j, k \le n$ and suppose $j \sim k_a$ and $k \sim k_b$. Then since $e(k_a, k_b)t_{k_b} \le t_{k_a}$,

$$\begin{aligned}
e(j,k)t_k &= e(j,k)e(k,k_b)t_{k_b} \\
&= e(j,k)e(k,k_b)e(k_a,k_b)^{-1}e(k_a,k_b)t_{k_b} \\
&\le e(j,k)e(k,k_b)e(k_a,k_b)^{-1}t_{k_a} \\
&= e(j,k)e(k,k_b)e(k_a,k_b)^{-1}e(j,k_a)^{-1}e(j,k_a)t_{k_a} \\
&= e(j,k)e(k,k_b)e(k_a,k_b)^{-1}e(j,k_a)^{-1}t_j \le t_j
\end{aligned}$$

since $e(j,k_a)e(k_a,k_b) = \frac{e(k_a,k_b)}{e(k_a,j)} \le \frac{e(k_a,j)e(j,k_b)}{e(k_a,j)} \le e(j,k)e(k,k_b)$. It follows that $e(j,k)t_k \le t_j$ as required, so $\sigma(\mathbf{t}^\flat) \in \Gamma(\mathbf{E})$. Thus if $\mathbf{t} = (t_1, \ldots, t_n) \in \Gamma(\mathbf{E})$, the coordinates t_j for $j \notin \{k_1, \ldots, k_s\}$ are uniquely determined by $\Pi(\mathbf{t})$. It follows that the mapping $\Pi : \Gamma(\mathbf{E}) \to \Gamma(\mathbf{E}^\flat)$ is one-to-one and onto, with the inverse mapping given by σ. We have established the following.

LEMMA 14.9. *Let $\mathbf{E}$ be an $n \times n$ matrix satisfying the basic hypothesis (2.10). Let $j \sim k$ if and only if $1 = e(j,k)e(k,j)$. This is an equivalence relation, and suppose there are s distinct equivalence classes. Let $\{k_1, \ldots, k_s\}$ be representatives of the distinct equivalence classes, and let $\mathbf{E}^\flat$ be the $s \times s$ reduced matrix whose entries are $\{e(k_a, k_b) : 1 \le a, b \le s\}$.*

a) *The reduced matrix $\mathbf{E}^\flat$ satisfies the basic hypothesis (2.10).*
b) *$1 < e(k_a,k_b)e(k_b,k_a)$ for all $1 \le a \ne b \le s$.*
c) *The dimension of $\Gamma(\mathbf{E})$ is s.*
d) *$\Gamma(\mathbf{E}^\flat) \subset \mathbb{R}^s$ has dimension s, and thus has non-empty interior.*
e) *Let $\Pi : \mathbb{R}^n \to \mathbb{R}^s$ be given by $\Pi(\mathbf{t}) = \Pi(t_1, \ldots, t_n) = (t_{k_1}, \ldots, t_{k_s})$. Then $\Pi : \Gamma(\mathbf{E}) \to \Gamma(\mathbf{E}^\flat)$ is one-to-one and onto.*
f) *Let $V = \Big\{\mathbf{t} = (t_1, \ldots, t_n) \in \mathbb{R}^n : t_j = e(j,k_a)t_{k_a}$ if $j \sim k_a\Big\}$. Then V is an s-dimensional subspace of $\mathbb{R}^n$, and $\Gamma(\mathbf{E}) \subset V$.*

CHAPTER 15

Appendix II: Estimates for homogeneous norms

Consider a family of dilations on $\mathbb{R}^m$ given by $\lambda \cdot \mathbf{x} = (\lambda^{d_1} x_1, \ldots, \lambda^{d_m} x_m)$ with homogeneous dimension $Q = d_1 + \cdots + d_m$. Let $N : \mathbb{R}^m \to [0, \infty)$ be a smooth homogeneous norm relative to this family so that $N(\mathbf{x}) \approx |x_1|^{1/d_1} + \cdots + |x_m|^{1/d_m}$. For any $\boldsymbol{\gamma} = (\gamma_1, \ldots, \gamma_m) \in \mathbb{N}^m$ let $[\![\gamma]\!] = d_1\gamma_1 + \cdots + d_m\gamma_m$.

PROPOSITION 15.1. *Let $\rho(\mathbf{x})$ be homogeneous of degree p. For $\boldsymbol{\gamma} \in \mathbb{N}^m$ there is a constant $C_\gamma > 0$ so that $|\partial_{\mathbf{x}}^{\gamma} \rho(\mathbf{x})| \leq C_\gamma\, N(\mathbf{x})^{p - [\![\gamma]\!]}$.*

PROOF. Let $\boldsymbol{\gamma} = (\gamma_1, \ldots, \gamma_m) \in \mathbb{N}^m$. Differentiating the equation $\lambda^p\, \rho(\mathbf{x}) = \rho(\lambda \cdot \mathbf{x})$ we obtain

$$\lambda^p\, \partial^\gamma \rho(\mathbf{x}) = \lambda^{\sum_{j=1}^m d_j\gamma_j} \partial^\gamma \rho(\lambda \cdot \mathbf{x}) = \lambda^{[\![\gamma]\!]} \partial^\gamma \rho(\lambda \cdot \mathbf{x}).$$

It follows that $|\partial_{\mathbf{x}}^{\gamma} \rho(\mathbf{x})|^{1/(p - [\![\gamma]\!])}$ is homogeneous of degree 1. Since N is continuous and strictly positive on the compact set $\{\mathbf{x} : N(\mathbf{x}) = 1\}$, it follows that $|\partial_{\mathbf{x}}^{\gamma} \rho(\mathbf{x})| \leq C_\gamma N(\mathbf{x})^{p - [\![\gamma]\!]}$. □

PROPOSITION 15.2. *There is a constant c so that $\big|\{\mathbf{x} \in \mathbb{R}^m : N(\mathbf{x}) \leq A\}\big| = c\, A^Q$.*

PROOF. Let $\mathbb{B}_m(1) = \{\mathbf{x} \in \mathbb{R}^m : N(\mathbf{x}) < 1\}$. Then $N(\mathbf{x}) \leq A$ if and only if $N(A^{-1} \cdot \mathbf{x}) = A^{-1} N(\mathbf{x}) \leq 1$, and this is true if and only if $A^{-1} \cdot \mathbf{x} \in \mathbb{B}_N(1)$. If $T_A(\mathbf{x}) = A \cdot \mathbf{x}$, this means that $\mathbf{x} \in T_A(\mathbb{B}(1))$. Since $\det(T_A) = A^Q$, the Proposition follows. □

PROPOSITION 15.3. *If $\mathbf{a} = (a_1, \ldots, a_m) \in \mathbb{N}^m$, write $\mathbf{x}^{\mathbf{a}} = \prod_{j=1}^m x_j^{a_j}$.*

(a) *For $\mathbf{a} \in \mathbb{N}^m$ and $M < [\![a]\!] + Q$ there exists $C > 0$ so that*

$$\int_{N(\mathbf{x}) \leq A} |\mathbf{x}^{\mathbf{a}}| N(\mathbf{x})^{-M}\, d\mathbf{x} \leq C\, A^{[\![\mathbf{a}]\!] + Q - M}.$$

(b) *For $\mathbf{a} \in \mathbb{N}^m$, $M > [\![a]\!] + Q$ there exists $C > 0$ so that*

$$\int_{N(\mathbf{x}) \geq A} |\mathbf{x}^{\mathbf{a}}| N(\mathbf{x})^{-M}\, d\mathbf{x} \leq C\, A^{[\![\mathbf{a}]\!] + Q - M}.$$

(c) *For $\mathbf{a} \in \mathbb{N}^m$, $M = [\![a]\!] + Q$, and $B > 4A$, there exists $C > 0$ so that*

$$\int_{A < N(\mathbf{x}) < B} |\mathbf{x}^{\mathbf{a}}| N(\mathbf{x})^{-M}\, d\mathbf{x} \leq C\, \log\left[\frac{B}{A}\right].$$

PROOF. We have $|\mathbf{x}^{\mathbf{a}}| = \prod_{j=1}^{m} |x_j|^{a_j} \lesssim \prod_{j=1}^{m} N(\mathbf{x})^{a_j d_j} = N(\mathbf{x})^{[\![\mathbf{a}]\!]}$. Thus if $[\![\mathbf{a}]\!] + Q - M > 0$,

$$\begin{aligned}\int_{N(\mathbf{x})\le A} |\mathbf{x}^{\mathbf{a}}| N(\mathbf{x})^{-M}\, d\mathbf{x} &\lesssim \sum_{j=1}^{\infty} \int_{2^{-j}A<N(\mathbf{x})\le 2^{-j+1}A} N(\mathbf{x})^{[\![\mathbf{a}]\!]-M}\, d\mathbf{x} \\ &\approx \sum_{j=1}^{\infty} (2^{-j}A)^{[\![\mathbf{a}]\!]-M} \big|\{\mathbf{x} : N(\mathbf{x}) \le 2^{-j+1}A\}\big| \\ &\approx \sum_{j=1}^{\infty} (2^{-j}A)^{[\![\mathbf{a}]\!]-M} (2^{-j+1}A)^{Q} \lesssim A^{[\![\mathbf{a}]\!]+Q-M}.\end{aligned}$$

Similarly, if $[\![\mathbf{a}]\!] + Q - M < 0$,

$$\begin{aligned}\int_{N(\mathbf{x})\ge A} |\mathbf{x}^{\mathbf{a}}| N(\mathbf{x})^{-M}\, d\mathbf{x} &\lesssim \sum_{j=0}^{\infty} \int_{2^{j}A\le N(\mathbf{x})<2^{j+1}A} N(\mathbf{x})^{[\![\mathbf{a}]\!]-M}\, d\mathbf{x} \\ &\approx \sum_{j=0}^{\infty} (2^{j}A)^{[\![\mathbf{a}]\!]-M} \big|\{\mathbf{x} : N(\mathbf{x}) \le 2^{-j+1}A\}\big| \\ &\lesssim \sum_{j=0}^{\infty} (2^{j}A)^{[\![\mathbf{a}]\!]-M} (2^{-j}A)^{Q} \\ &\lesssim A^{[\![\mathbf{a}]\!]+Q-M}.\end{aligned}$$

Finally, if $[\![\mathbf{a}]\!] + Q - M = 0$,

$$\int_{A\le N(\mathbf{x})\le B} |\mathbf{x}^{\mathbf{a}}| N(\mathbf{x})^{-M}\, d\mathbf{x} \lesssim \int_{A\le N(\mathbf{x})\le B} N(\mathbf{x})^{[\![\mathbf{a}]\!]-M}\, d\mathbf{x}.$$

If $B > 4A$, choose $j < k$ so that $2^{j-1} < A \le 2^{j} \le 2^{k} \le B < 2^{k+1}$. Then

$$\begin{aligned}\int_{A\le N(\mathbf{x})\le B} N(\mathbf{x})^{[\![\mathbf{a}]\!]-M}\, d\mathbf{x} &\le \sum_{l=j-1}^{k+1} \int_{2^{l}\le N(\mathbf{x})<2^{l+1}} N(\mathbf{x})^{[\![\mathbf{a}]\!]-M}\, d\mathbf{x} \\ &\lesssim \sum_{l=j-1}^{k+1} 2^{l([\![\mathbf{a}]\!]-M)} \big|\{\mathbf{x} : N(\mathbf{x}) \le 2^{l+1}\}\big| \\ &\lesssim \sum_{l=j-1}^{k+1} 2^{l([\![\mathbf{a}]\!]-M+Q)} = k - j + 2 \approx \log\left[\frac{B}{A}\right].\end{aligned}$$

This completes the proof. □

PROPOSITION 15.4. Let $m \in \mathcal{C}^{\infty}(\mathbb{R}^m \setminus \{0\})$ have compact support, and let $L = \{l_1, \ldots, l_r\} \subset \{1, \ldots, m\}$. Suppose for every $\gamma \in \mathbb{N}$ there is a constant $C > 0$ so that if $j \in L$ then

$$\big|\partial_{\xi_j}^{\gamma} m(\xi_1, \ldots, \xi_m)\big| \le C N(\boldsymbol{\xi})^{-d_j \gamma}.$$

Put $K(\mathbf{x}) = \int_{\mathbb{R}^m} e^{2\pi i\langle \mathbf{x},\boldsymbol{\xi}\rangle} m(\boldsymbol{\xi})\, d\boldsymbol{\xi}$. There is a constant $C > 0$ independent of the support of m so that

$$\big|K(\mathbf{x})\big| \le C\,\big(\sum_{j\in L} |x_j|^{1/d_j}\big)^{-Q}.$$

PROOF. Let $\varphi \in \mathcal{C}_0^\infty(\mathbb{R})$ with $\varphi(t) \equiv 1$ if $|t| \le 1$, $\varphi(t) \equiv 0$ if $|t| \ge 2$ and $0 \le \varphi(t) \le 1$ for all t. We have

$$\begin{aligned} K(\mathbf{x}) &= \int_{\mathbb{R}^m} e^{2\pi i\langle \mathbf{x},\boldsymbol{\xi}\rangle} \varphi\big(\lambda N(\boldsymbol{\xi})\big) m(\boldsymbol{\xi})\, d\boldsymbol{\xi} + \int_{\mathbb{R}^m} e^{2\pi i\langle \mathbf{x},\boldsymbol{\xi}\rangle} \big[1 - \varphi\big(\lambda N(\boldsymbol{\xi})\big)\big] m(\boldsymbol{\xi})\, d\boldsymbol{\xi} \\ &= I + II. \end{aligned}$$

Since m is bounded, we have

$$\big|I\big| \le ||m||_\infty \big|\{\boldsymbol{\xi} : N(\boldsymbol{\xi}) \le 2\lambda^{-1}\}\big| \lesssim \lambda^{-Q}.$$

On the other hand, the chain, the product rule, and Proposition 15.1 show that

$$\big|\partial_{\xi_j}^{\gamma} \big[1 - \varphi\big(\lambda N(\boldsymbol{\xi})\big)\big] m(\boldsymbol{\xi})\big| \le C N(\boldsymbol{\xi})^{-d_j\gamma}$$

since any derivative of $\varphi(\lambda N(\boldsymbol{\xi}))$ is supported where $\lambda N(\boldsymbol{\xi}) \le 2$. For $j \in L$ we can integrate by parts M times in the variable ξ_j and obtain

$$\begin{aligned} \big|II\big| &\lesssim |x_j|^{-M} \int_{N(\boldsymbol{\xi})\ge\lambda^{-1}} \Big|\partial_{\xi_j}^{M}\Big[\big[1 - \varphi\big(\lambda N(\boldsymbol{\xi})\big)\big] m(\boldsymbol{\xi})\Big]\Big|\, d\boldsymbol{\xi} \\ &\lesssim |x_j|^{-M} \int_{N(\boldsymbol{\xi})\ge\lambda^{-1}} N(\boldsymbol{\xi})^{-Md_j}\, d\boldsymbol{\xi} \\ &\lesssim |x_j|^{-M} \lambda^{-Q+d_jM}. \end{aligned}$$

Setting $\lambda = |x_j|^{1/d_j}$, it follows that $|K(\mathbf{x})| \lesssim |x_j|^{-Q/d_j}$ for every $j \in L$, and hence that

$$|K(\mathbf{x})| \lesssim \min_{j\in L} |x_j|^{-Q/d_j} \approx \Big(\sum_{j\in L} |x_j|^{1/d_j}\Big)^{-Q}.$$

This completes the proof. $\square$

REMARK 15.5. Note that in this Proposition, we only assume estimates on "pure" ξ_j derivatives of m for $j \in L$. The conclusion is that the inverse Fourier transform $K = \check{m}$ has the expected decay relative to the partial norm $N_L(\mathbf{x}) = \sum_{j\in L} |x_j|^{1/d_j}$. If $L = \{1, \ldots, n\}$, the conclusion is that $K(\mathbf{x}) \lesssim N(\mathbf{x})^{-Q}$.

CHAPTER 16

Appendix III: Estimates for geometric sums

PROPOSITION 16.1. Let $\epsilon > 0$, $a_1, \ldots, a_m > 0$ with $m \geq 2$ and $\prod_{j=1}^m a_j \neq 1$. Set

$$\Gamma = \Big\{ I = (i_1, \ldots, i_m) \in \mathbb{Z}^m : a_j i_j \leq i_{j+1} < 0 \text{ for } 1 \leq j \leq m-1,\ a_m i_m \leq i_1 < 0 \Big\}.$$

For $I = (i_1, \ldots, i_m) \in \Gamma$ set $\Lambda(I) = [i_2 - a_1 i_1] + \cdots + [i_m - a_{m-1} i_{m-1}] + [i_1 - a_m i_m]$. Then there is a constant $C > 0$ depending on $\{\epsilon, a_1, \ldots, a_m\}$ so that $\sum_{I \in \Gamma} 2^{-\epsilon \Lambda(I)} \leq C$.

PROOF. Γ is a convex cone and $\{\mathbf{t} \in \Gamma : \sum_{j=1}^m |t_j| = 1\}$ is compact. Then $\Lambda(\mathbf{t}) \geq 0$ on Γ, and if $\Lambda(\mathbf{t}) = 0$ with $\mathbf{t} \in \Gamma$, then $t_1 = a_m t_m$ and $t_{j+1} = a_j t_j$ for $1 \leq j \leq m-1$, so that $t_1 = \big(\prod_{j=1}^m a_j\big) t_1$. Since $\prod_{j=1}^m a_j \neq 1$, it follows that $\Lambda(\mathbf{t}) > 0$ for $t \in \Gamma$. In particular, by compactness there exists $\eta > 0$ so that if $\mathbf{t} \in \Gamma$ and $|\mathbf{t}| = 1$ then $\Lambda(\mathbf{t}) > \eta$. Since Λ is continuous, there is an open neighborhood of $\{\mathbf{t} \in \Gamma : \sum_{j=1}^m |t_j| = 1\}$ on which $\Lambda(\mathbf{t}) > \eta$. By homogeneity it follows that $\Lambda(\mathbf{t}) \geq \eta \sum_{j=1}^m |t_j|$ for all$\mathbf{t} \in \Gamma$.

If $I \in \Gamma$, then $\Lambda(I) \geq \eta \sum_{j=1}^n |i_j|$. Thus we have

$$\sum_{I \in \Gamma} 2^{-\epsilon \Lambda(I)} \leq \sum_{I \in \Gamma} 2^{-\epsilon \eta \sum_{j=1}^m |i_j|} \leq \Big[\sum_{i=0}^{\infty} 2^{-\epsilon \eta i}\Big]^m$$

which is a constant depending only on ϵ and $a_1, \ldots, a_m$. This completes the proof. □

Now let $\mathbf{E} = \{e(j,k)\}$ be an $n \times n$ matrix satisfying (2.10) such that $1 < e(j,k)e(k,j)$ for all $1 \leq j,k \leq n$. Let $\epsilon > 0$. Let $\tau : \{1, \ldots, n\} \to \{1, \ldots, n\}$ with $\tau(j) \neq j$ for all j. Let $0 \leq b < a \leq n$. Put

$$\begin{aligned}
\Gamma_{\mathbb{Z}}(\mathbf{E}) &= \Big\{ I = (i_1, \ldots, i_n) \in \mathbb{Z}^n : e(j,k) i_k \leq i_j, \qquad 1 \leq j,k \leq n \Big\},\\
\Gamma_{\mathbb{Z}}(\mathbf{E}') &= \Big\{ I' = (i_1, \ldots, i_a) \in \mathbb{Z}^a : e(j,k) i_k \leq i_j, \qquad 1 \leq j,k \leq a \Big\},\\
\Gamma_{\mathbb{Z}}(\mathbf{E}'') &= \Big\{ I'' = (i_{a+1}, \ldots, i_n) \in \mathbb{Z}^{n-a} : e(j,k) i_k \leq i_j, \quad a+1 \leq j,k \leq n \Big\}.
\end{aligned}$$

Note that $\Gamma_{\mathbb{Z}}(\mathbf{E}) \subset \Gamma_{\mathbb{Z}}(\mathbf{E}') \times \Gamma_{\mathbb{Z}}(\mathbf{E}'')$. Also, if $I'' \in \Gamma_{\mathbb{Z}}(\mathbf{E}'')$ then

$$\Lambda(I'') = \Big\{ I' = (i_1, \ldots, i_a) \in \mathbb{Z}^a : (I', I'') \in \Gamma_{\mathbb{Z}}(\mathbf{E}) \Big\} \subset \Gamma_{\mathbb{Z}}(\mathbf{E}').$$

PROPOSITION 16.2. Let $R_{b+1}, \ldots, R_a$ be positive real numbers. Then

$$F = \sum_{I' \in \Gamma_{\mathbb{Z}}(\mathbf{E}')} \Big[\prod_{j=1}^{b} 2^{-\epsilon [i_j - e(j,\tau(j)) i_{\tau(j)}]}\Big] \Big[\prod_{j=b+1}^{a} \min\Big\{(2^{i_j} R_j)^{+\epsilon}, (2^{i_j} R_j)^{-\epsilon}\Big\}\Big]$$

converges and is bounded by a constant C, depending on $\epsilon, \mathbf{E}$ but independent of $R_{b+1}, \ldots, R_a$. Here it is understood that if $b=0$ the first product is empty, and if $0<b=a$ the second product is empty.

PROOF. We argue by induction on $n \geq 1$. When $n=1$ we have

$$F \leq \sum_{j\in\mathbb{Z}} \min\Big\{(2^jR)^{+\epsilon}, (2^jR)^{-\epsilon}\Big\}.$$

which converges and is independent of R. Thus assume that the result is true for some $n-1\geq 1$. There are two possibilities. First suppose that $b=0$. Then

$$\begin{aligned}
F &= \sum_{I'\in\Gamma_{\mathbb{Z}}(\mathbf{E}')} \prod_{j=1}^{a} \min\Big\{(2^{i_j}R_j)^{+\epsilon}, (2^{i_j}R_j)^{-\epsilon}\Big\} \\
&\leq \sum_{(i_1,\ldots,i_a)\in\mathbb{Z}^a} \prod_{k=1}^{a} \min\Big\{(2^{i_k}R_k)^{+\epsilon}, (2^{i_k}R_k)^{-\epsilon}\Big\} \\
&= \prod_{j=1}^{a}\Bigg[\sum_{i_j\in\mathbb{Z}} \min\Big\{(2^{i_j}R_j)^{+\epsilon}, (2^{i_j}R_j)^{-\epsilon}\Big\}\Bigg]
\end{aligned}$$

which is independent of $R_1, \ldots, R_a$.

Next suppose that $b\geq 1$ so that the product $\prod_{j=1}^{b} 2^{-\epsilon\left[i_j-e(j,\tau(j))i_{\tau(j)}\right]}$ is non-empty. There are now two possibilities.

Case 1:

Suppose there exists $j_0\in\{1,\ldots,b\}$ such that $j_0\neq\tau(k)$ for any $k\in\{1,\ldots,b\}$. Without loss of generality, assume $j_0=1$. Set

$$\widetilde{\Gamma}_{\mathbb{Z}}(\mathbf{E}') = \Big\{\widetilde{I}' = (i_2,\ldots,i_a)\in\mathbb{Z}^{a-1} : e(j,k)i_k \leq i_j, \quad 2\leq j,k\leq a\Big\}.$$

Then $\Gamma_{\mathbb{Z}}(\mathbf{E}') \subset \Big\{(i_1,\widetilde{I}') : e(1,\tau(1))i_{\tau(1)} \leq i_1 \text{ and } \widetilde{I}'\in\widetilde{\Gamma}_{\mathbb{Z}}(\mathbf{E}')\Big\}$. Since the index i_1 does not appear in the product $\prod_{j=2}^{b} 2^{-\epsilon\left[i_j-e(j,\tau(j))i_{\tau(j)}\right]}$, the sum F is dominated by

$$\begin{aligned}
&\sum_{\widetilde{I}'\in\widetilde{\Gamma}_{\mathbb{Z}}(\mathbf{E}')}\Bigg[\prod_{j=2}^{b} 2^{-\epsilon\left[i_j-e(j,\tau(j))i_{\tau(j)}\right]}\Bigg]\Bigg[\prod_{j=b+1}^{a} \min\Big\{(2^{i_j}R_j)^{+\epsilon}, (2^{i_j}R_j)^{-\epsilon}\Big\}\Bigg] \\
&\qquad\qquad\times\Bigg[\sum_{\substack{i_1\in\mathbb{Z}\\ e(1,\tau(1))i_{\tau(1)}\leq i_1}} 2^{-\epsilon\left[i_1-e(1,\tau(1))i_{\tau(1)}\right]}\Bigg]
\end{aligned}$$

But

$$\sum_{\substack{i_1\in\mathbb{Z}\\ e(1,\tau(1))i_{\tau(1)}\leq i_1}} 2^{-\epsilon\left[i_1-e(1,\tau(1))i_{\tau(1)}\right]} \leq \Big[1-2^{-\epsilon e(1,\tau(1))}\Big]^{-1},$$

and so

$$F \le \frac{1}{1 - 2^{-\epsilon e(1,\tau(1))}} \sum_{\widetilde{I}' \in \widetilde{\Gamma}_{\mathbb{Z}}(\mathbf{E}')} \left[\prod_{j=2}^{b} 2^{-\epsilon\left[i_j - e(j,\tau(j)) i_{\tau(j)}\right]} \right] \left[\prod_{j=b+1}^{a} \min\left\{ (2^{i_j} R_j)^{+\epsilon}, (2^{i_j} R_j)^{-\epsilon} \right\} \right].$$

The result now follows by induction since in the remaining sum we have replaced $\{1, \ldots, n\}$ with $\{2, \ldots, n\}$.

Case 2:

Now assume that each $j \in \{1, \ldots, b\}$ is equal to $\tau(k)$ for at least one $k \in \{1, \ldots, b\}$. Let $\sigma(j) \in \{1, \ldots, b\}$ be a choice of some pre-image of j. Then $\sigma : \{1, \ldots, b\} \to \{1, \ldots, b\}$, and since $\tau(\sigma(j)) = j$, it follows that σ is one-to-one. Thus σ (and hence also τ) is a permutation of $\{1, \ldots, b\}$. It follows that we can decompose $\{1, \ldots, b\}$ into a disjoint union of non-empty subsets $C_1, \ldots, C_p$ which are the minimal closed orbits of the mapping τ. We can write $C_l = \{j_{l,1}, \ldots, j_{l,m_l}\}$ with $\tau(j_{l,k}) = j_{l,k+1}$ for $1 \le k \le m_l - 1$ and $\tau(j_{l,k_l}) = j_{l,1}$. Then for $1 \le l \le p$ let

$$\Gamma_l' = \left\{ (i_{l,1}, \ldots, i_{l,m_l}) \in \mathbb{Z}^{m_l} : e(j_{l,k}, \tau(j_{l,k})) i_{\tau(j_{l,k})} \le i_{j_{l,k}} \; 1 \le k \le m_l \right\}.$$

Then $\Gamma_{\mathbb{Z}}(\mathbf{E}') \subset \Gamma_1' \times \cdots \times \Gamma_p$, and so

$$F \le \prod_{l=1}^{p} \Big[\sum_{I_l \in \Gamma_l'} \prod_{k=1}^{m_l} 2^{-\epsilon[i_{j_{l,k}} - e(j_{l,k}, \tau(j_{l,k})) i_{\tau(j_{l,k})}]} \Big] \times \prod_{k=a+1}^{b} \Big[\sum_{i_k \in \mathbb{Z}} \min\left\{ (2^{i_k} R_k)^{+\epsilon}, (2^{i_k} R_k)^{-\epsilon} \right\} \Big]$$

However $\sum_{I_l \in \Gamma_l'} \prod_{k=1}^{m_l} 2^{-\epsilon[i_{j_{l,k}} - e(j_{l,k}, \tau(j_{l,k})) i_{\tau(j_{l,k})}]}$ is uniformly bounded by Proposition 14.8 and Proposition 16.1. Also, $\sum_{i_k \in \mathbb{Z}} \min\left\{ (2^{i_k} R_k)^{+\epsilon}, (2^{i_k} R_k)^{-\epsilon} \right\}$ is bounded independently of R_k. Thus F is bounded independently of i_{s+1} and $R_{a+1}, \ldots, R_b$. This completes the proof. □

PROPOSITION 16.3. Let $\mathbf{E} = \{e(j,k)\}$ be an $n \times n$ matrix satisfying (2.10), and suppose that $1 < e(j,k) e(k,j)$ for all $j \neq k$. Let $\mathbf{e} = \max\{e(j,k) : 1 \le j, k \le n\}$. Let $\alpha_j > 0$, $M_j \ge 0$ for $1 \le j \le n$. If $M \ge \sum_{j=1}^{n} (\mathbf{e}(\alpha_j + 1) + M_j)$ there is a constant $C = C(M_1, \ldots, M_n, n, \mathbf{e})$ so that for all $(A_1, \ldots, A_n)$ with $A_j > 0$,

$$\sum_{I \in \Gamma_{\mathbb{Z}}(\mathbf{E})} \Big[\prod_{j=1}^{n} 2^{-i_j \alpha_j} \Big] \Big[1 + \sum_{k=1}^{n} 2^{-i_k} A_k \Big]^{-M} \le C \prod_{j=1}^{n} \Big[A_1^{e(j,1)} + \cdots + A_n^{e(j,n)} \Big]^{-\alpha_j} \prod_{j=1}^{n} (1 + A_j)^{-M_j}.$$

PROOF. Let $N_j(A) = \sum_{k=1}^n A_k^{e(j,k)}$. If $I = (i_1, \ldots, i_n) \in \Gamma_{\mathbb{Z}}(\mathbf{E})$ then for any $1 \leq j \leq n$ we have $-i_j < -e(j,k)i_k$ and so

$$\begin{aligned} 1 + 2^{-i_j} N_j(A) &\leq 1 + \sum_{k=1}^n 2^{-i_k e(j,k)} A_k^{e(j,k)} \\ &\leq 1 + \sum_{k=1}^n \left[2^{-i_k} A_k\right]^{e(j,k)} \\ &\lesssim \left[1 + \sum_{k=1}^n 2^{-i_k} A_k\right]^{\mathbf{e}}. \end{aligned}$$

Therefore

$$\begin{aligned} \left[1 + \sum_{k=1}^n 2^{-i_k} A_k\right]^{-M} &\leq \prod_{j=1}^n \left[1 + \sum_{k=1}^n 2^{-i_k} A_k\right]^{-\mathbf{e}(\alpha_j+1)} \prod_{j=1}^n \left[1 + A_j\right]^{-M_j} \\ &\lesssim \prod_{j=1}^n \left[1 + 2^{-i_j} N_j(A)\right]^{-(\alpha_j+1)} \prod_{j=1}^n \left[1 + A_j\right]^{-M_j}, \end{aligned}$$

and so

$$\begin{aligned} &\sum_{I \in \Gamma_{\mathbb{Z}}(\mathbf{E})} \left[\prod_{j=1}^n 2^{-i_j \alpha_j}\right] \left[1 + \sum_{k=1}^n 2^{-i_k} A_k\right]^{-M} \\ &\qquad \lesssim \sum_{I \in \Gamma_{\mathbb{Z}}(\mathbf{E})} \prod_{j=1}^n 2^{-i_j \alpha_j} \left[1 + 2^{-i_j} N_j(A)\right]^{-(\alpha_j+1)} \prod_{j=1}^n \left[1 + A_j\right]^{-M_j} \\ &\qquad \lesssim C \prod_{j=1}^n \left[\sum_{i_j \leq 0} 2^{-i_j \alpha_j} \left[1 + 2^{-i_j} N_j(A)\right]^{-\alpha_j - 1}\right] \prod_{j=1}^n \left[1 + A_j\right]^{-M_j} \\ &\qquad \lesssim \prod_{j=1}^n N_j(A)^{-\alpha_j} \prod_{j=1}^n \left[1 + A_j\right]^{-M_j}. \end{aligned}$$

This completes the proof. □

Bibliography

[CGP13] T. C. Collins, A. Greenleaf, and M. Pramanik, *A multi-dimensional resolution of singularities with applications to analysis*, Amer. J. Math. **135** (2013), no. 5, 1179–1252, DOI 10.1353/ajm.2013.0042. MR3117305

[DLM10] Y. Ding, G. Z. Lu, and B. L. Ma, *Multi-parameter Triebel-Lizorkin and Besov spaces associated with flag singular integrals*, Acta Math. Sin. (Engl. Ser.) **26** (2010), no. 4, 603–620, DOI 10.1007/s10114-010-8352-8. MR2591640

[FS82] G. B. Folland and E. M. Stein, *Hardy spaces on homogeneous groups*, Mathematical Notes, vol. 28, Princeton University Press, Princeton, N.J.; University of Tokyo Press, Tokyo, 1982. MR657581

[Gło10a] P. Głowacki, *Composition and L^2-boundedness of flag kernels*, Colloq. Math. **118** (2010), no. 2, 581–585, DOI 10.4064/cm118-2-13. MR2602167

[Gło10b] P. Głowacki, *Correction to "Composition and L^2-boundedness of flag kernels" [MR2602167]*, Colloq. Math. **120** (2010), no. 2, 331, DOI 10.4064/cm120-2-12. MR2679042

[Gło13] P. Głowacki, *L^p-boundedness of flag kernels on homogeneous groups via symbolic calculus*, J. Lie Theory **23** (2013), no. 4, 953–977. MR3185206

[GS77] P. C. Greiner and E. M. Stein, *Estimates for the $\overline{\partial}$-Neumann problem*, Princeton University Press, Princeton, N.J., 1977. Mathematical Notes, No. 19. MR0499319

[HL10] Y. Han and G. Lu, *Some recent works on multiparameter Hardy space theory and discrete Littlewood-Paley analysis*, Trends in partial differential equations, Adv. Lect. Math. (ALM), vol. 10, Int. Press, Somerville, MA, 2010, pp. 99–191. MR2648281

[HLL10] Y. Han, J. Li, and G. Lu, *Duality of multiparameter Hardy spaces H^p on spaces of homogeneous type*, Ann. Sc. Norm. Super. Pisa Cl. Sci. (5) **9** (2010), no. 4, 645–685. MR2789471

[HLL13] Y. Han, J. Li, and G. Lu, *Multiparameter Hardy space theory on Carnot-Carathéodory spaces and product spaces of homogeneous type*, Trans. Amer. Math. Soc. **365** (2013), no. 1, 319–360, DOI 10.1090/S0002-9947-2012-05638-8. MR2984061

[HLS14] Y. Han, G. Lu, and E. Sawyer, *Flag Hardy spaces and Marcinkiewicz multipliers on the Heisenberg group*, Anal. PDE **7** (2014), no. 7, 1465–1534, DOI 10.2140/apde.2014.7.1465. MR3293443

[LZ13] G. Z. Lu and Y. P. Zhu, *Singular integrals and weighted Triebel-Lizorkin and Besov spaces of arbitrary number of parameters*, Acta Math. Sin. (Engl. Ser.) **29** (2013), no. 1, 39–52, DOI 10.1007/s10114-012-1402-7. MR3001008

[MPR15] D. Müller, M. M. Peloso, and F. Ricci, *Analysis of the Hodge Laplacian on the Heisenberg group*, Mem. Amer. Math. Soc. **233** (2015), no. 1095, vi+91, DOI 10.1090/memo/1095. MR3289035

[MRS95] D. Müller, F. Ricci, and E. M. Stein, *Marcinkiewicz multipliers and multi-parameter structure on Heisenberg (-type) groups. I*, Invent. Math. **119** (1995), no. 2, 199–233, DOI 10.1007/BF01245180. MR1312498

[NRS01] A. Nagel, F. Ricci, and E. M. Stein, *Singular integrals with flag kernels and analysis on quadratic CR manifolds*, J. Funct. Anal. **181** (2001), no. 1, 29–118, DOI 10.1006/jfan.2000.3714. MR1818111

[NRSW12] A. Nagel, F. Ricci, E. Stein, and S. Wainger, *Singular integrals with flag kernels on homogeneous groups, I*, Rev. Mat. Iberoam. **28** (2012), no. 3, 631–722, DOI 10.4171/RMI/688. MR2949616

[PS82] D. H. Phong and E. M. Stein, *Some further classes of pseudodifferential and singular-integral operators arising in boundary value problems. I. Composition of operators*, Amer. J. Math. **104** (1982), no. 1, 141–172, DOI 10.2307/2374071. MR648484

[RS88] F. Ricci and E. M. Stein, *Harmonic analysis on nilpotent groups and singular integrals. II. Singular kernels supported on submanifolds*, J. Funct. Anal. **78** (1988), no. 1, 56–84, DOI 10.1016/0022-1236(88)90132-2. MR937632

[Rua10] Z. Ruan, *Weighted Hardy spaces in the three-parameter case*, J. Math. Anal. Appl. **367** (2010), no. 2, 625–639, DOI 10.1016/j.jmaa.2010.02.010. MR2607286

[Rua11] Z. Ruan, *Non-isotropic flag singular integrals on multi-parameter Hardy spaces*, Taiwanese J. Math. **15** (2011), no. 2, 473–499. MR2810164

[Ste93] E. M. Stein, *Harmonic analysis: real-variable methods, orthogonality, and oscillatory integrals*, Princeton Mathematical Series, vol. 43, Princeton University Press, Princeton, NJ, 1993. With the assistance of Timothy S. Murphy; Monographs in Harmonic Analysis, III. MR1232192

[SY13] E. M. Stein and P.-L. Yung, *Pseudodifferential operators of mixed type adapted to distributions of k-planes*, Math. Res. Lett. **20** (2013), no. 6, 1183–1208, DOI 10.4310/MRL.2013.v20.n6.a15. MR3228630

[Str14] B. Street, *Multi-parameter singular integrals*, Annals of Mathematics Studies, vol. 189, Princeton University Press, Princeton, NJ, 2014. MR3241740

[WL12] X. Wu and Z. Liu, *Characterizations of multiparameter Besov and Triebel-Lizorkin spaces associated with flag singular integrals*, J. Funct. Spaces Appl., posted on 2012, Art. ID 275791, 18, DOI 10.1155/2012/275791. MR2923798

[Wu14a] X. Wu, *An atomic decomposition characterization of flag Hardy spaces $H^p_F(\mathbb{R}^n \times \mathbb{R}^m)$ with applications*, J. Geom. Anal. **24** (2014), no. 2, 613–626, DOI 10.1007/s12220-012-9347-8. MR3192289

[Wu14b] X. Wu, *Weighted norm inequalities for flag singular integrals on homogeneous groups*, Taiwanese J. Math. **18** (2014), no. 2, 357–369, DOI 10.11650/tjm.18.2014.3667. MR3188508

[Yan09] D. Yang, *Besov and Triebel-Lizorkin spaces related to singular integrals with flag kernels*, Rev. Mat. Complut. **22** (2009), no. 1, 253–302, DOI 10.5209/rev_REMA.2009.v22.n1.16358. MR2499336

Index of Symbols

Editorial Information

To be published in the *Memoirs*, a paper must be correct, new, nontrivial, and significant. Further, it must be well written and of interest to a substantial number of mathematicians. Piecemeal results, such as an inconclusive step toward an unproved major theorem or a minor variation on a known result, are in general not acceptable for publication.

Papers appearing in *Memoirs* are generally at least 80 and not more than 200 published pages in length. Papers less than 80 or more than 200 published pages require the approval of the Managing Editor of the Transactions/Memoirs Editorial Board. Published pages are the same size as those generated in the style files provided for $\mathcal{A}\mathcal{M}\mathcal{S}$-LaTeX or $\mathcal{A}\mathcal{M}\mathcal{S}$-TeX.

Information on the backlog for this journal can be found on the AMS website starting from `http://www.ams.org/memo`.

A Consent to Publish is required before we can begin processing your paper. After a paper is accepted for publication, the Providence office will send a Consent to Publish and Copyright Agreement to all authors of the paper. By submitting a paper to the *Memoirs*, authors certify that the results have not been submitted to nor are they under consideration for publication by another journal, conference proceedings, or similar publication.

Information for Authors

Memoirs is an author-prepared publication. Once formatted for print and on-line publication, articles will be published as is with the addition of AMS-prepared frontmatter and backmatter. Articles are not copyedited; however, confirmation copy will be sent to the authors.

Initial submission. The AMS uses Centralized Manuscript Processing for initial submissions. Authors should submit a PDF file using the Initial Manuscript Submission form found at `www.ams.org/submission/memo`, or send one copy of the manuscript to the following address: Centralized Manuscript Processing, MEMOIRS OF THE AMS, 201 Charles Street, Providence, RI 02904-2294 USA. If a paper copy is being forwarded to the AMS, indicate that it is for *Memoirs* and include the name of the corresponding author, contact information such as email address or mailing address, and the name of an appropriate Editor to review the paper (see the list of Editors below).

The paper must contain a *descriptive title* and an *abstract* that summarizes the article in language suitable for workers in the general field (algebra, analysis, etc.). The *descriptive title* should be short, but informative; useless or vague phrases such as "some remarks about" or "concerning" should be avoided. The *abstract* should be at least one complete sentence, and at most 300 words. Included with the footnotes to the paper should be the 2010 *Mathematics Subject Classification* representing the primary and secondary subjects of the article. The classifications are accessible from `www.ams.org/msc/`. The Mathematics Subject Classification footnote may be followed by a list of *key words and phrases* describing the subject matter of the article and taken from it. Journal abbreviations used in bibliographies are listed in the latest *Mathematical Reviews* annual index. The series abbreviations are also accessible from `www.ams.org/msnhtml/serials.pdf`. To help in preparing and verifying references, the AMS offers MR Lookup, a Reference Tool for Linking, at `www.ams.org/mrlookup/`.

Electronically prepared manuscripts. The AMS encourages electronically prepared manuscripts, with a strong preference for $\mathcal{A}\mathcal{M}\mathcal{S}$-LaTeX. To this end, the Society has prepared $\mathcal{A}\mathcal{M}\mathcal{S}$-LaTeX author packages for each AMS publication. Author packages include instructions for preparing electronic manuscripts, samples, and a style file that generates the particular design specifications of that publication series. Though $\mathcal{A}\mathcal{M}\mathcal{S}$-LaTeX is the highly preferred format of TeX, author packages are also available in $\mathcal{A}\mathcal{M}\mathcal{S}$-TeX.

Authors may retrieve an author package for *Memoirs of the AMS* from `www.ams.org/journals/memo/memoauthorpac.html`. The *AMS Author Handbook* is available in PDF format from the author package link. The author package can also be obtained free of charge by sending email to tech-support@ams.org or from the Publication Division, American Mathematical Society, 201 Charles St., Providence, RI 02904-2294, USA. When requesting an author package, please specify $\mathcal{A}_\mathcal{M}\mathcal{S}$-LaTeX or $\mathcal{A}_\mathcal{M}\mathcal{S}$-TeX and the publication in which your paper will appear. Please be sure to include your complete mailing address.

After acceptance. The source files for the final version of the electronic manuscript should be sent to the Providence office immediately after the paper has been accepted for publication. The author should also submit a PDF of the final version of the paper to the editor, who will forward a copy to the Providence office.

Accepted electronically prepared files can be submitted via the web at www.ams.org/submit-book-journal/, sent via FTP, or sent on CD to the Electronic Prepress Department, American Mathematical Society, 201 Charles Street, Providence, RI 02904-2294 USA. TeX source files and graphic files can be transferred over the Internet by FTP to the Internet node ftp.ams.org (130.44.1.100). When sending a manuscript electronically via CD, please be sure to include a message indicating that the paper is for the *Memoirs*.

Electronic graphics. Comprehensive instructions on preparing graphics are available at www.ams.org/authors/journals.html. A few of the major requirements are given here.

Submit files for graphics as EPS (Encapsulated PostScript) files. This includes graphics originated via a graphics application as well as scanned photographs or other computer-generated images. If this is not possible, TIFF files are acceptable as long as they can be opened in Adobe Photoshop or Illustrator.

Authors using graphics packages for the creation of electronic art should also avoid the use of any lines thinner than 0.5 points in width. Many graphics packages allow the user to specify a "hairline" for a very thin line. Hairlines often look acceptable when proofed on a typical laser printer. However, when produced on a high-resolution laser imagesetter, hairlines become nearly invisible and will be lost entirely in the final printing process.

Screens should be set to values between 15% and 85%. Screens which fall outside of this range are too light or too dark to print correctly. Variations of screens within a graphic should be no less than 10%.

Any graphics created in color will be rendered in grayscale for the printed version unless color printing is authorized by the Managing Editor and the Publisher. In general, color graphics will appear in color in the online version.

Inquiries. Any inquiries concerning a paper that has been accepted for publication should be sent to memo-query@ams.org or directly to the Electronic Prepress Department, American Mathematical Society, 201 Charles St., Providence, RI 02904-2294 USA.

Editors

This journal is designed particularly for long research papers, normally at least 80 pages in length, and groups of cognate papers in pure and applied mathematics. Papers intended for publication in the *Memoirs* should be addressed to one of the following editors. The AMS uses Centralized Manuscript Processing for initial submissions to AMS journals. Authors should follow instructions listed on the Initial Submission page found at www.ams.org/memo/memosubmit.html.

SELECTED PUBLISHED TITLES IN THIS SERIES

1224 **Werner Hoffmann and Satoshi Wakatsuki,** On the Geometric Side of the Arthur Trace Formula for the Symplectic Group of Rank 2, 2018

1223 **Sergey Fomin and Professor Dylan Thurston,** Cluster Algebras and Triangulated Surfaces Part II: Lambda Lengths, 2018

1222 **Maurice Duits and Kurt Johansson,** On Mesoscopic Equilibrium for Linear Statistics in Dyson's Brownian Motion, 2018

1221 **Francesco Lin,** A Morse-Bott Approach to Monopole Floer Homology and the Triangulation Conjecture, 2018

1220 **Paata Ivanisvili, Dmitriy M. Stolyarov, Vasily I. Vasyunin, and Pavel B. Zatitskiy,** Bellman Function for Extremal Problems in BMO II: Evolution, 2018

1219 **Olivier Frécon,** Algebraic $\overline{\mathbb{Q}}$-Groups as Abstract Groups, 2018

1218 **Gabriella Pinzari,** Perihelia Reduction and Global Kolmogorov Tori in the Planetary Problem, 2018

1217 **Chin-Yu Hsiao,** Szegő Kernel Asymptotics for High Power of CR Line Bundles and Kodaira Embedding Theorems on CR Manifolds, 2018

1216 **Robert Lipshitz, Peter Ozsváth, and Dylan P. Thurston,** Bordered Heegaard Floer Homology, 2018

1215 **Lior Fishman, David Simmons, and Mariusz Urbański,** Diophantine Approximation and the Geometry of Limit Sets in Gromov Hyperbolic Metric Spaces, 2018

1214 **David Chataur, Martintxo Saralegi-Aranguren, and Daniel Tanré,** Intersection Cohomology, Simplicial Blow-Up and Rational Homotopy, 2018

1213 **Sebastiano Carpi, Yasuyuki Kawahigashi, Roberto Longo, and Mihály Weiner,** From Vertex Operator Algebras to Conformal Nets and Back, 2018

1212 **Roelof Bruggeman, Youngju Choie, and Nikolaos Diamantis,** Holomorphic Automorphic Forms and Cohomology, 2018

1211 **Shouhei Honda,** Elliptic PDEs on Compact Ricci Limit Spaces and Applications, 2018

1210 **Zhou Gang, Dan Knopf, and Israel Michael Sigal,** Neckpinch Dynamics for Asymmetric Surfaces Evolving by Mean Curvature Flow, 2018

1209 **Cristian Anghel, Iustin Coandă, and Nicolae Manolache,** Globally Generated Vector Bundles with Small c_1 on Projective Spaces, 2018

1208 **Matthew Harrison-Trainor,** Degree Spectra of Relations on a Cone, 2018

1207 **Alastair J. Litterick,** On Non-Generic Finite Subgroups of Exceptional Algebraic Groups, 2018

1206 **Anne-Laure Dalibard and Laure Saint-Raymond,** Mathematical Study of Degenerate Boundary Layers: A Large Scale Ocean Circulation Problem, 2018

1205 **Charles Collot,** Type II Blow Up Manifolds for the Energy Supercritical Semilinear Wave Equation, 2018

1204 **Rui Palma,** Crossed Products by Hecke Pairs, 2018

1203 **Xiao Xiong, Quanhua Xu, and Zhi Yin,** Sobolev, Besov and Triebel-Lizorkin Spaces on Quantum Tori, 2018

1202 **Naiara V. de Paulo and Pedro A. S. Salomão,** Systems of Transversal Sections Near Critical Energy Levels of Hamiltonian Systems in $\mathbb{R}^4$, 2018

1201 **Bernhelm Booß-Bavnbek and Chaofeng Zhu,** The Maslov Index in Symplectic Banach Spaces, 2018

1200 **Francis Nier,** Boundary Conditions and Subelliptic Estimates for Geometric Kramers-Fokker-Planck Operators on Manifolds with Boundaries, 2018